Cambridge Elements

Elements in Organizational Response to Climate Change
edited by
Aseem Prakash
University of Washington
Jennifer Hadden
University of Maryland
David Konisky
Indiana University
Matthew Potoski
UC Santa Barbara

REBEL GOVERNANCE IN THE AGE OF CLIMATE CHANGE

Kathleen Gallagher Cunningham
University of Maryland

Leonardo Gentil-Fernandes
University of Tennessee

Elisabeth Gilmore
Carleton University

Reyko Huang
Texas A&M University

Danielle F. Jung
Emory University

Cyanne E. Loyle
Pennsylvania State University
Peace Research Institute Oslo

CAMBRIDGE
UNIVERSITY PRESS

Shaftesbury Road, Cambridge CB2 8EA, United Kingdom

One Liberty Plaza, 20th Floor, New York, NY 10006, USA

477 Williamstown Road, Port Melbourne, VIC 3207, Australia

314–321, 3rd Floor, Plot 3, Splendor Forum, Jasola District Centre,
New Delhi – 110025, India

103 Penang Road, #05-06/07, Visioncrest Commercial, Singapore 238467

Cambridge University Press is part of Cambridge University Press & Assessment,
a department of the University of Cambridge.

We share the University's mission to contribute to society through the pursuit of
education, learning and research at the highest international levels of excellence.

www.cambridge.org
Information on this title: www.cambridge.org/9781009569446
DOI: 10.1017/9781009569439

First published 2025

A catalogue record for this publication is available from the British Library

ISBN 978-1-009-56944-6 Hardback
ISBN 978-1-009-56946-0 Paperback
ISSN 2753-9342 (online)
ISSN 2753-9334 (print)

Additional resources for this publication at www.cambridge.org/ORCC_Cunningham

Rebel Governance in the Age of Climate Change

Elements in Organizational Response to Climate Change

DOI: 10.1017/9781009569439
First published online: May 2025

Kathleen Gallagher Cunningham
University of Maryland

Leonardo Gentil-Fernandes
University of Tennessee

Elisabeth Gilmore
Carleton University

Reyko Huang
Texas A&M University

Danielle F. Jung
Emory University

Cyanne E. Loyle
Pennsylvania State University
Peace Research Institute Oslo

Author for correspondence: Kathleen Gallagher Cunningham,
kgcunnin@umd.edu

Abstract: In many areas experiencing severe impacts from climate change, it is not the state, but rather rebel groups who wield authority over populations. Rebels are often engaged in responding and adapting to the risks and impacts of climate change as part of their local governance efforts; however, a systematic consideration of the activities and implications has been lacking. This Element looks at a set of behaviors we call "rebel environmental governance" (REG+). This refers to rebel actions aimed at protecting or managing the natural environment to affect civilian welfare amidst increasing pressures of climate change. A framework is advanced for understanding why rebels engage in environmental governance and the implications for security and climate governance. The Element brings rebel organizations into the conversation on climate change, highlighting their role in areas where state power is contested, weak, or absent. This title is also available as Open Access on Cambridge Core.

This Element also has a video abstract: www.cambridge.org/ORCC-Cunningham

Keywords: conflict, climate change, environment, rebellion, security

ISBNs: 9781009569446 (HB), 9781009569460 (PB), 9781009569439 (OC)
ISSNs: 2753-9342 (online), 2753-9334 (print)

Contents

An online appendix for this publication can be accessed at
https://www.cambridge.org/ORCC-Cunningham

1 Introduction

The Ogaden National Liberation Front affirms that we shall confront all initiatives, which negatively impact our environment as a matter of national duty to protect our environment for future generations.

— ONLF[1]

Today humanity is confronting this great and serious climatic threat.
— *Secretary General of Hezbollah in 2010*[2]

Like the ONLF in Ethiopia, armed groups around the globe have undertaken a variety of actions to govern over issues relating to the physical environment, with potential implications for climate change. Like Hezbollah in Lebanon, armed groups have also begun to speak directly about climate change and its challenges, and in some cases have proclaimed their environmental actions as necessary responses to climate change. Studies of rebel governance often miss that rebel groups are governing in this space, and studies of climate change responses often neglect that armed non-state actors are important players in environmental and climate governance, both directly and as a second-order effect of their behavior.[3] In this Element, we aim to introduce rebel groups to the set of actors considered relevant for governance around issues of climate change. We use the terms rebel group, armed actors, and armed non-state actors interchangeably throughout, all of which capture non-state organizations that employ violence against the state for political goals. We provide a novel look, theoretically and empirically, at environmental governance by rebels, with potential effects for climate mitigation and adaptation. Even if rebel groups are not thinking about climate change, the structures of environmental governance they establish and/or enforce have second-order effects on climate policy.

For some, it may seem improbable that rebel groups would engage in environmental governance based on their average capabilities, their often confrontational stance towards the international community, and above all their use of violence to upend the prevailing political order.[4] Armed conflicts are themselves highly destructive events with devastating consequences on societies, livelihoods, and the natural environment.[5] Yet, this Element demonstrates that rebel groups

---

[1] From the ONLF Political Program (ONLF, n.d.).

[2] Quoted in Karagiannis (2015, p. 186).

[3] Armed actors have been clearly identified as key actors impacting vulnerability (Buhaug & Von Uexkull, 2021).

[4] Rebel groups are by definition "illegal" in the states in which they operate. However, rebel actors have been increasingly engaging with areas of international governance typically seen as the purview of recognized states, including human rights and international law (Jo, 2015).

[5] The detrimental impacts of conflict have been documented in a number of contexts (Fergusson et al., 2014; GebreMichael et al., 1992; Hendrix & Glaser, 2011).

engage in a broad range of actions that are within the scope of our understanding of environmental and climate governance, including direct management of natural resources, forward planning for the risks of environmental and climate-induced hazards, and responses to climate impacts – and that rebel environmental governance can and does take place alongside, and in response to, the devastations of armed conflict.

1.1 Rebel Environmental Governance in Rojava, Syria

On December 26, 2015 Syrian Democratic Forces (SDF) troops, in cooperation with U.S. and coalition forces, captured the Tishreen Dam on the Euphrates River from the Islamic State (IS). This was a turning point in the SDF's fight against the Islamic State, but also in the ability of the SDF to govern northeast Syria, which was pivotal to maintaining its de facto autonomy from the Syrian government. The dam is the keystone infrastructure project in the water governance of the region as its reservoir feeds the domestic and agricultural water needs of northeast Syria. Furthermore, it provides electricity to large areas of major cities such as Aleppo, Manbij, Kobani, and Raqqah. Since its capture, the SDF government has administered the Tishreen Dam with the goal to provide water for the region's populations in the face of worsening climate conditions and deteriorating infrastructure (Minttu, 2023; Sary, 2016).

The SDF is a coalition rebel group composed of several armed actors that has been fighting against the Syrian government and IS. The most dominant group in the coalition, the Democratic Union Party (PYD), is a Kurdish political party formed in Syria in 2004 with assistance from the Turkey-based Kurdistan Workers' Party (PKK). The coalition advocates for Kurdish self-governance through a "Democratic Confederalism Model" based on the works of Abdullah Öcalan and Murray Bookchin, both of which emphasize ecological sustainability as a core tenet of their ideology (Barkhoda, 2016; Cemgil & Hoffmann, 2016). The SDF's campaign for self-determination has been a part of a wider and devastating war: As of 2024, the Uppsala Conflict Data Program (UCDP) estimates that the conflict in Syria has caused over 407,475 battle-related deaths, and the United Nations High Commissioner for Refugees (UNHCR) estimates over 7 million internally displaced people and 5 million refugees (UNHCR, 2024).

Upon taking control of the Rojava region, the SDF established a decentralized government with local-level assemblies that oversee all aspects of local governance. The local assemblies have committees charged with governance over health, education, and defense. Many local assemblies have also created agriculture and ecology committees (Hatahet, 2019; Knapp & Jongerden, 2016). At the federal level, the SDF has several committees involved in environmental management and disaster response, including the Oil and Natural Resources Office, the Development and Planning Office, the Office of Humanitarian Affairs, the Economy and Agriculture Commission, and the Health and Environment Commissions (Hatahet, 2019).

These institutions oversee several climate change-related governance initiatives and reflect the SDF's ideology of social ecology.[6] Their charges include devising adaptation strategies such as diversifying agricultural production and promoting less water-intensive and organic agriculture (Barkhoda, 2016; Cemgil & Hoffmann, 2016; Hatahet, 2019; Jongerden, 2010). Some of these programs, including banning chemical fertilizers and incentivizing the use of organic ones, are explicitly aimed at restoring and preventing further deterioration of water and soil quality in the region (Barkhoda, 2016).[7] Efforts have also been made to increase the water security of Northeast Syria, especially in the face of chronic water shortages due to climate change (Jongerden, 2010; Sottimano & Samman, 2022). The SDF cooperates with a broad range of actors to address the impact of climate change on livelihoods, including NGOs such as Blumont and Action Against Hunger (ACF), as well as the Syrian government.[8] Despite its central role as an armed group in the Syrian civil war, the SDF has become a leading actor in environmental and climate governance and especially in the coordination of climate adaptation in the region.

1.2 Climate Governance beyond the State

Climate-induced hazards have been increasing in both frequency and intensity and are projected to worsen in the next decades, even under stringent climate mitigation policy (IPCC, 2022). The adequacy and appropriateness of responses to these impacts have recently been recognized as having the potential to be risk factors in their own right (Simpson et al., 2021). Understanding of climate adaptation globally (Berrang-Ford et al., 2019) and in conflict-affected

---

[6] The writings of Murray Bookchin are the primary source of inspiration for the ideological leader of the PYD, Abdullah Öcalan, and the group's view of the role of ecology in their governance project (Barkhoda, 2016). Bookchin's work on social ecology broadly viewed a transition to societies and markets fundamentally structured around small-scale and communal social relations not dependent on property or production for exchange as both a social good and a necessity to integrate societies with their environmental surroundings which he argued was needed to survive the climate crisis (Biehl, 2014).

[7] This deterioration was largely due to existing practices but that these are increasingly exacerbated by climate change.

[8] On NGO cooperation, see program websites (Blumont, n.d.; *Middle East | Action Against Hunger – Global Hunger Relief,* n.d.). Burcu Özçelik describes how the PYD and the Assad regime maintain several parallel institutions and continue to pay salaries to several Government employees including those involved in health and infrastructure (areas largely responsible for climate-related disaster relief and climate adaptation) who are involved in SDF governance (Özçelik, 2020).

countries in particular (Sitati et al., 2021), however, is limited. Much of the attention from academics and practitioners focuses on government responses at the national level (such as analysis of the politics surrounding the 2016 Paris Climate Agreement) or on the implementation of climate agreements and treaties and is predominantly focused on climate mitigation. For adaptation, while scholarship has addressed individual, household, and community responses to climate change challenges, questions remain about how to link locally defined needs and responses with larger internationalized policy frameworks (Persson, 2019), especially in ways that could incorporate armed actors.

Climate discourse often acknowledges the importance of entities such as corporations (Yadin, 2023), nongovernmental organizations (NGOs) (Hadden & Bush, 2021), and subnational state institutions (Hale, 2018). Other actors, from traditional authorities and community leaders to armed groups and criminal organizations, remain notably absent from public debate on climate change.[9] Studies have highlighted the hybrid and multilateral nature of climate governance that integrates non-state actors (Bäckstrand et al., 2017; Kuyper et al., 2018). However, these approaches tend to emphasize positive cooperation among diverse entities, masking the reality of competition – and sometimes open conflict – between various actors for authority, control, and legitimacy.

As the SDF example in Syria highlights, rebel groups that violently challenge the state are often central actors in both local governance and crisis management in times of armed conflict. In 2023 alone, there were fifty-nine active armed conflicts which killed an estimated 122,518 people (Davies et al., 2024). Despite their violent strategies, rebel groups engage in local governance in numerous conflict-affected contexts. To varying degrees, they provide social services such as healthcare and schooling, establish local order through the introduction of their own laws and courts, organize communities through youth associations and women's cooperatives, and encourage self-sustenance through agricultural production and land reform – even as they wield violence to fight state forces, and even as some engage in civilian targeting and terrorist tactics in doing so.[10] While the effectiveness of, and the degree of coercion used in, these governance measures vary, existing scholarship largely depicts rebel governance as a kind

---

[9] Studies have looked at the potential moderating effects of local institutions that include resource use rules on violent conflict (Linke et al., 2018).

[10] A growing literature addresses rebels as governing actors in a number of contexts (Arjona et al., 2015; Breslawski, 2021; Cunningham et al., 2021; Cunningham & Loyle, 2021; Driscoll, 2015; Flanigan, 2008; Florea, 2020; Heger & Jung, 2017; Huang, 2016b; Huang & Sullivan, 2021; Huddleston & Hall, 2023; Jackson, 2018; Kasfir et al., 2017; Ledwidge, 2017; Loyle, 2021; Loyle et al., 2023; Malejacq, 2019; Mampilly, 2011; Martínez & Eng, 2018; Parkinson, 2013; Podder, 2017; Revkin, 2020; Rubin, 2020; Schoon, 2017; Seymour, 2017; Stewart, 2018; Terpstra, 2020; Worrall, 2018).

of wartime social contract between rebel rulers and local communities. Rebel groups provide order, goods, and services to civilians in return for their loyalty, "taxes," intelligence, logistical help, and other contributions to the war effort. Additionally, as actors aspiring to political power, rebel groups govern in order to gain legitimacy – they seek local and international recognition of their right and ability to rule.

In contexts as diverse as the Philippines, Somalia, and Morocco, rebel groups have exercised governance authority around environmental issues in their areas of operation. For example, the Patriotic Union of Kurdistan (PUK) has often controlled the Ministry of Agriculture and Irrigation in Iraqi Kurdistan, primarily tasked with managing the region's water resources to achieve food sustainability. This area has seen substantial drought conditions, impacting water scarcity and deforestation (Eklund et al., 2017). In Colombia, since 2022 the rebel group Estado Mayor Central (EMC) has instituted logging bans in regions under its control, with enforcement via fines for violations, thus creating parallel efforts to curb deforestation to those of the Colombian Government (Collins, 2023). These two examples demonstrate some of the ways armed actors engage in environmental governance and the potential spillover effects rebel actions can have for climate adaptation and mitigation efforts.[11]

Despite the robust literature on rebel governance, there is currently little recognition of the fact that rebel groups engage in environmental governance and are often highly attuned to the effects of climate change, and correspondingly, limited understanding of the implications of these rebels' actions for climate mitigation and adaptation. We build on existing knowledge of rebel governance to introduce, theorize, and empirically examine environmental governance by rebel groups.

1.3 Defining Rebel Environmental Governance

This Element centers on the idea that rebel groups engage in governance over the environment, and that this behavior should be examined in the context of climate change – rebel environmental action can affect broader climate governance efforts, and rebel groups themselves often attribute their behavior to the impacts and risks of climate change. Our effort marks a clear departure from prevailing approaches to understanding rebels and the environment. Existing scholarship predominantly focuses on processes that degrade the environment, whether through rebel exploitation of natural resources for profit (Conrad et al., 2019; Lujala, 2010; Marks, 2019; Weinstein, 2006) or direct rebel attacks on the

---

[11] The impact of rebel behavior on climate mitigation is frequently indirect, such as forest protection's impact on carbon storage, which in turn impacts global climate.

environment (Feuer, 2023), or they address environment- and climate-induced hazards as a "threat multiplier" which facilitates rebel violence and induces conflict (Abrahams, 2020; Asaka, 2021; Hendrix & Salehyan, 2012; Ide, 2023). As such, our baseline for understanding rebels and the environment is that rebels exploit and degrade the natural environment both directly and indirectly, and that conflict itself causes significant harm to the environment. The purpose of this Element is to expand this baseline understanding to better reflect broader realities: Rebel actions can harm the environment, but rebels also actively manage the environment and even take steps to protect it.

In this section, we explain what we mean by "rebel environmental govern-ance" in the Element. We further discuss how this relates to conceptualizations of "climate governance," acknowledging that these conceptual definitions are debated in numerous literatures and are further complicated when armed non-state entities are the implementers in question.[12]

Governance refers to the process of making and enforcing rules over a specific population or policy area.[13] Rebel groups are armed non-state actors who challenge the state violently in order to seek concessions from the state or to directly gain power, whether through regime overthrow or secession.[14] They govern in many ways, over a variety of different issue areas and with different degrees of institutional and procedural formality. Rebel groups often do so in order to gain resources, increased authority, and strategic advantages in conflict. This Element investigates rebel governance over issues related to the natural environment, such as rebel control over water access and management of land use, as well as rebel governance over climate hazards and their impacts, such as disaster management. In this study, we use the term "rebel environmental governance (REG+)" to refer to rebel governance aimed at managing the natural environment. We focus specifically, though not exclusively, on governance activities most directly related to civilian welfare in the context of climate change (including current impacts and future risks).

We take rebel interactions with the environment as our starting point for introducing rebel groups as relevant actors in the climate space. The "+" reflects our understanding that some of the activities we include under the umbrella of "environmental governance" more directly concern social programs, such as those that have been identified as important for building resilience and

---

[12] See Okereke et al. (2009).

[13] Behaviorally, scholars understand governance as consisting of rule-making, rule enforcement, and the provision of social services; see Loyle et al. (2022).

[14] Our definition follows from the Uppsala Conflict Data Program (Davies et al., 2023; N. P. Gleditsch et al., 2002). In practice, rebel groups seek a wide variety of types of accommodations from the states they challenge. All of these demands, however, can be grouped into claims about either more local power or more power at the center of the state.

managing climate impacts. For example, managing human displacement and migration related to environmental hazard, or bureaucratic efforts (such as creating a "ministry" of environmental affairs) are activities we include that are not direct management of the natural environment per se. Through these "+" activities, we bridge several aspects of environmental and climate governance. We acknowledge climate governance is a broad category of actions (related to mitigation and adaptation) that include negotiations, decisions, and activities at multiple levels, from local to global, aimed at addressing climate change and its impacts, and that those actions are more than responses to and about the natural environment. Nevertheless, the basic premise of environmental governance is the effort to exercise authority over the management of the natural environment and its impacts on communities. This most closely aligns with activities classified as adaptation, although conservation may have positive impacts for climate mitigation as well.

We note several key features of this definition that shape our approach to REG+. First, we make no assumptions about the intent of rebel groups in addressing and adapting to climate change. We are interested in rebel environmental governance regardless of whether the rebels themselves point to climate change as driving their behavior, and hence rebels need not explicitly describe their environmental governance as a response to climate change for it to constitute a REG+ action in this study. Second, we are agnostic as to the legality of the behavior related to environmental governance. Rebels could be engaged in illicit activity (such as the Taliban encouraging poppy cultivation) that fit broadly into the category of agricultural management, which is within our scope of environmental governance. Third, we do not presume that the outcome of rebel environmental governance must be effective, positively impacting the natural environment.[15] However, we do assume that the behavior is not designed to harm the environment (such as the use of environmental warfare (Feuer, 2023)).[16] Fourth, rebel environmental governance can occur as part of hybrid governance. Rebel efforts may be taken in coordination with the state, NGOs, or other actors. This is especially relevant to our definition, as environmental governance may include facilitating access for other actors to provide on-the-ground environmental services. Fifth, we do not require that all changes in the environment related to the activities in our database are necessarily directly attributed to climate change (c.f. https://www.worldweather attribution.org/) and include some activities related only to the environment or

---

[15] This work also intersects with the experimentation literature on climate governance (Kivimaa et al., 2017).

[16] Likewise, we would not consider direct extraction (mining, logging) as "environmental governance," but rather natural resource extraction separate from governing. Some scholarship has included "high value agriculture" as natural resource extraction (Walsh et al., 2018).

natural hazard events. Imposing the condition of attribution of the hazard rather than types of events that are associated with climate change would place an unreasonable condition for inclusion in the dataset. Additionally, climate change may be indirectly affecting many natural hazards in ways that were not previously documented or where there is uncertainty.[17] Finally, our effort to document and provide a framework for understanding rebel environmental governance is in no way intended to legitimize rebel groups or endorse their behavior, or to minimize their acts of violence and coercion. As is standard in rebel governance studies, our aim is a scholarly engagement with an observed phenomenon, with potential implications for our understanding of civilian welfare, conflict dynamics, security, and climate governance.

Regarding issues of intent, as addressed earlier we do not assume rebel groups are managing environmental issues in direct or acknowledged response to climate change, though they may well be – and indeed, many do use the language of climate change, as our data will show. Rather, because of the challenges of inferring motivation from observed behavior, we place no requirements that rebel governance over the environment and associated hazards be linked explicitly to climate change. Instead, we focus on observable instances where rebel groups engage in governance related to the environment. In the age of climate change, we believe there is value in documenting that rebel groups introduce changes to local agricultural practices, take steps to protect forests, cooperate with state authorities to regulate water access, or invite humanitarian organizations to address the immediate impacts of drought or storms, whether or not the group in question describe such behavior as a response to climate change.

We acknowledge that our conceptualization of REG+ is in tension with some conventional definitions of climate governance. In conventional discourse, global climate mitigation generally concerns international cooperation and arrangements that aim to reduce greenhouse gas (GHG) emissions, primarily under the auspices of the United Nations Framework Convention on Climate Change (UNFCCC) and the country-level negotiations at the Convention of the Parties (COPs) to the UNFCCC. However, climate governance's wide remit also involves a range of non-state actors and the multiple ways they interact with global governance and more local structures (Okereke et al., 2009; R. B. Stewart et al., 2013); indeed, the 2015 Paris Agreement has more formally given a role to non-state actors in achieving climate goals (Kuyper et al., 2018). This shift to acknowledging and

---

[17] The dataset is well documented and any given user could opt to remove certain records depending on their exclusion criteria.

integrating non-state actors in our understanding of climate governance is, at least in part, due to the challenges[18] experienced in international climate negotiations (Gupta, 2016), which have led to a more diverse set of actors participating in climate governance and increasing interest in its effectiveness (Jordan et al., 2015).[19] The need for engagement at multiple levels and across multiple policy spaces is even more pronounced for climate adaptation, where there are often urgent needs at local levels and greater disconnection from international regimes addressing climate change (Biesbroek & Lesnikowski, 2018; Persson, 2019).

As the first investigation into the environmental governance activities of rebel groups around the world, the breadth of our definition allows us to be inclusive in our understanding of how rebel groups engage with environmental issues, thus facilitating the broadest possible theorizing of this underexplored phenomenon.

1.4 Expanding Our Understanding of Climate Governance to Include REG+

A foremost goal of this Element is to broaden the discourse on environmental governance to more fully and accurately reflect the wide range of actors that play a role in addressing climate-related challenges in conflict-affected areas. Existing literature on the relationship between armed conflict and climate change depicts rebel groups almost exclusively as producers of violence. Indeed, a range of empirical studies identify a circle of "violence, vulnerability, and climate change" (Buhaug & Von Uexkull, 2021).[20] This Element draws attention to impacts beyond violence (and its consequences) that rebels have on the areas in which they operate and the populations with whom they engage. We do so by integrating rebel groups into the broader context of multilayered governance around environmental issues – contexts which variously include local authorities, domestic and international NGOs, and civic groups. Figure 1 uses a graphic representation of the cyclic nature of conflict and climate change

---

[18] These challenges relate to a changing view of the "problem" presented by climate change, which has evolved from being seen as structured and technological, to less structured and political, to unstructured and ideological (Gupta, 2016).

[19] This perspective was adopted in the IPCC AR6 where it is recognized that climate governance occurs in the context of other key priorities, such as energy security and urban development, and recognizing that this engages with a large number of actors, including nongovernmental organizations, business groups, think tanks, trade unions, private governance arrangements, transnational networks, and substate authorities, all who can and do shape climate governance (Dubash, 2021). This is reflected in the growing literature in diverse governance realms, notably urban and local climate governance (Van Der Heijden, 2019) and national climate governance (Dubash, 2021).

[20] For a comprehensive survey of the climate-conflict connection see Koubi (2019).

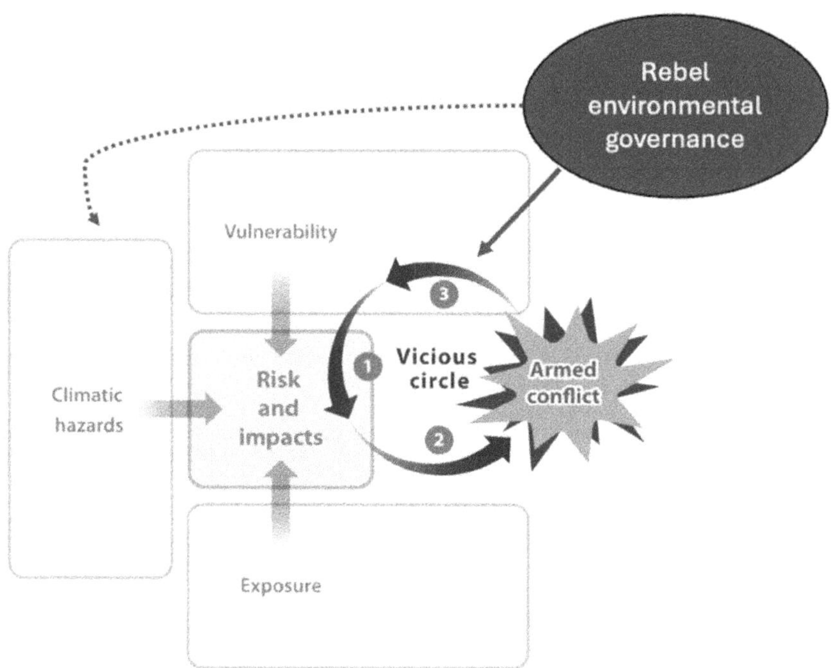

Figure 1 Rebel environmental governance and the climate-conflict cycle

impacts from Buhaug and Von Uexkull (2021) to show how rebel governance over the environment can affect these dynamics. In the original figure, Buhaug and Von Uexkull (2021) highlight how greater vulnerability increases the risks and impacts of climatic hazards (1), how these effects, in turn, increase the chance of armed conflict (2), and how conflict has a detrimental effect on vulnerability to future hazards (3).

Environmental governance by rebels intersects with this larger puzzle of conflict and climate change. REG+ comprises some climate mitigation activities, namely through land and forest conservation, and climate adaptation with its focus on reducing vulnerability. The dashed line indicates a first potential pathway of impact on climatic hazard via mitigation. While the scale of activity by rebel authorities at present may have quite limited impacts on the occurrence of climate hazards, some rebel actions (such as forest preservation as a carbon sink) can contribute to climate mitigation. Other activities such as establishing environment ministries and climate rhetoric may also show engagement with longer-term efforts to prevent increases in global warming levels. The solid line indicates a second pathway of impact on vulnerability. With more immediacy, rebels – as actors that shape citizen welfare (both positively and negatively) – play a critical role through climate adaptation related activities in shaping vulnerability to climate-induced hazards.

This project contributes not only to ongoing research on climate adaptation, mitigation, and disaster response, but also to scholarship on hybrid governance in conflict-affected states (Börzel & Risse, 2021; Loyle et al., 2023; Tapscott, 2023). As we discuss in the Element's conclusion, thinking about rebel groups as active agents in environmental governance opens new conversations around important academic and policy questions.

1.5 Mapping the Element

In this Element, we demonstrate that rebel groups around the world are engaging in governance and rhetoric related to the environment. Exploring this phenomenon, we address the central research question: How and under what conditions do rebel groups engage in governance over the natural environment? To preview, our framework proposes that rebel groups are political authorities who use military and political strategies to challenge state power and gain concessions. In this context, three interrelated concerns help explain rebel environmental governance. First, understanding that they must not only make military gains but also garner legitimacy locally and internationally, rebel groups engage in environmental governance to enhance their credibility and showcase their capacity as governors and as potentially viable state authorities. Second, as armed actors with goals to control territory, rebels seek to preserve the physical integrity of the very territories over which they lay political claim; the land itself holds material, political, and symbolic value and helps sustain the lives and livelihoods of its residents. Third, rebel groups govern over the environment for tactical reasons: features such as dense forests are mainstays of insurgent strategy as they provide physical cover and protection to fighters, while access to water is essential for their survival and operations and for livelihood within their popular bases of support. In short, environmental governance helps advance rebels' political, social, and tactical objectives.

We proceed in four sections. In Section 2, we engage with the broader literature on rebel governance, highlighting what we know about when and why rebels govern, and connect this with the state of knowledge about the relevance of armed non-state actors in the environment and climate change space. We then advance our framework and arguments in full, and distill a set of hypotheses from them. In Section 3, we present the REG+ data, which includes novel information on efforts by rebel groups to provide environmental services, respond to hazards, develop related institutions, and cooperate with other actors for all disputes over territorial claims active after 1989. In Section 4, we use REG+ and existing data on armed conflict to provide some limited tests of our theoretical implications centering on legitimacy-seeking behavior and local community investments. Furthermore, we provide eight case vignettes structured around our three empirical expectations both to illustrate our arguments and highlight the scope and complexity of rebel environmental governance.

In the concluding section (Section 5), we summarize the findings and lay out the implications of the Element for future research and policy. The recognition of rebel groups as relevant actors for our understanding of governance related to the environment and of efforts to address and respond to climate change raises a number of questions. We elaborate on research questions for rebel governance scholars, as well as scholarship on the environment and climate change. Finally, we engage with the difficult question of when and how it may be appropriate to work with violent non-state actors, centering on practical, political, and normative implications.

2 Rebel Groups as Governors

2.1 Why Do Rebel Groups Govern?

While armed conflicts involve violent confrontations between state and rebel forces, politically they are contests over the authority to govern territories and communities. States seek to retain their monopoly over the use of force, ability to provide basic services, and ensure public order; rebel groups aim to impair state rule and alter or supplant it with their own. Rebel governance can thus be understood as rebels' projection of authority over local populations through the provision of administrative and social services, aimed at shaping social and political life in accordance with their own visions. This fundamentally political contest over authority helps explain why states and rebels vie for control over taxation powers and the provision of electricity, schooling and health services in the midst of warfare, and why they both insist on the exclusive legitimacy of their own legal and judicial systems. In other words, rebel groups' campaigns against the state may be as much about winning the fight for authority and legitimacy in order to achieve their goals as about militarily defeating or imposing costs on the state.

Despite its volume, range of research questions, and some epistemological divergences, scholarship on rebel governance converges on the idea that rebel governance is a manifestation of the core struggle for governance powers between political contenders (Arjona et al., 2015; Duyvesteyn, 2017; Florea & Malejacq, 2023; Furlan, 2020; Heger & Jung, 2017; Huang, 2016a; Loyle et al., 2023, 2022; Mampilly, 2011; Staniland, 2012; Stewart, 2021; Worrall, 2018). Arguments cluster around three basic notions about why rebel groups govern. First, rebel governance establishes a system through which rebel groups can more readily source the physical, financial, and political assets necessary for their campaigns against the state. By serving as local governance providers and engaging with civilians, rebels can collect "taxes," gain valuable intelligence, demand other basic necessities, and, if successful, win local popular support and loyalty through the provision of welfare services and strategic messaging. Second, rebel governance can be explained by the rebels' quest for political authority and legitimacy. Both domestically and internationally, they wish to be acknowledged as capable power-holders and accepted as legitimate political entities rather than outlawed insurgents or a simple militia. Demonstrating governance and administrative capability, in their estimation, is one important way to ascend the ladder towards widespread recognition as legitimate authorities. Finally, rebel groups govern because in doing so they may gain a competitive edge against challengers. The latter include rival rebel groups against whom they compete for local control, resources, and popular support, as well as the state itself, as when rebel groups try to maximize concessions from the government in conflict negotiations.

The literature on rebel governance has made significant strides in explaining rebel politics in armed conflicts. Nevertheless, there remains a striking absence of focus on rebels' governance of the natural environment[21] and, in parallel, a lack of systematic data on how and to what extent rebel groups engage in environmental governance. Rebel environmental governance is poised to become an important area of study in light of the serious threats posed by climate change throughout conflict-affected societies (Barnett & Adger, 2007; Bergholt & Lujala, 2012; Buhaug et al., 2014; N. P. Gleditsch, 2012; Hsiang et al., 2013; Koubi, 2019; Koubi et al., 2012; Linke & Ruether, 2021; Von Uexkull & Buhaug, 2021). Thinking about rebel groups in the age of climate change propels us to broaden our contextual scope to recognize that threats in conflict-affected areas derive not only from instability and violence, but also from environmental degradation and hazards. Indeed, a senior official from Iraqi Kurdistan identified climate change as being on par with the Islamic State in terms of the region's top threats (Sen, 2023). Next, we demonstrate that while existing studies on rebel governance help explain its basic elements, rebel environmental governance also demands the generation of arguments that are unique to the phenomenon. The study of rebel environmental governance thus

---

[21] Numerous works examine rebels' control over natural resources such as oil, timber, and minerals, aimed at resource extraction and profiting from their trade (Conrad et al., 2019; Lujala, 2010; Marks, 2019; Weinstein, 2006). Our endeavor is distinct from such efforts: our concern is with rebel governance aimed at protecting the environment or managing the environment to affect civilian welfare, rather than the extractive potential of natural resources per se, though the environmental consequences of natural resource extraction would be within this study's analytical remit.

helps to not just affirm, but also advance, the study of rebel governance, and of conflict dynamics and environmental politics more broadly.

2.2 Governance Related to Climate Change

Outside of the literature on authority and governance in conflict settings, much attention has been devoted to state and international governance related to the challenges associated with climate change. Strong, supportive, and inclusive governance – along with finance and knowledge that support governance capacity – are seen as key enablers of climate action, and are associated with more ambitious plans and more effective implementation (IPCC, 2023b). Previously, we defined governance as the process of making and enforcing rules over a specific population or policy area. In the climate space, the Intergovernmental Panel on Climate Change (IPCC) elaborates this idea further by defining governance as "the structures, processes, and actions through which private and public actors interact to address societal goals. This includes formal and informal institutions and the associated norms, rules, laws, and procedures for deciding, managing, implementing, and monitoring policies and measures at any geographic or political scale, from global to local" (IPCC, 2023a, p. 164). In this context, climate governance relates to the efforts of mitigation – primarily through reducing greenhouse gas emissions and shifting land use patterns – and adaptation – activities that reduce the risks posed by climate change.

This conceptualization of climate governance, as well as its associated programming, takes international cooperation between states and national implementation efforts as the primary means through which the international community strives towards its climate commitments. In practice, however, a wide range of actors, including subnational governments, cities, NGOs, and civil society groups comprise a "complex regime" of climate governance (Dubash, 2021; Keohane & Victor, 2011). For instance, climate change mitigation involves legislation and frameworks to guide implementation by a wider range of actors, including civil society and the private sector (Sapiains et al., 2021). A polycentric form of governance is especially important for climate adaptation, where the public sector takes the lead in such areas as the provision of social safety nets and spatial planning, while communities and households undertake autonomous adaptation efforts (especially following extreme weather events) (Aerts et al., 2018). Adaptation has historically been understood as a local-level effort, but is now increasingly understood to span multiple levels of governance, including the involvement of non-state actors.[22]

---

[22] There has also been substantial consideration of what makes governance effective in the context of climate change (Keskitalo & Preston, 2019; Owen, 2020). Engagement with local groups, especially those who are most vulnerable to climate risks, has been shown to be most effective through structures and systems that emphasize participatory governance. Other decision-making

Climate action in conflict-affected areas is complicated by the dynamics of these settings, as existing studies have begun to acknowledge (Siddiqi, 2018). Some examples of disaster risk reduction activities involving armed non-state actors have been noted, such as in the Philippines for livelihood support for farmers (Walch, 2018) and Afghanistan for reforestation (Mena & Hilhorst, 2021). There are increasing signals of armed groups' engagement in climate-related adaptation (Jackson et al., 2023), interactions with climate-related activities (e.g. deforestation (Wrathall et al., 2020)), and even the desirability of their engagement in adaptation and disaster risk management (Raleigh et al., 2024). Nevertheless, climate governance involving armed groups in conflict-affected settings has not been widely documented (Sitati et al., 2021).

2.3 Rebel Group Provision of Environmental Governance

Why do rebel groups engage in environmental governance? Our approach focuses on rebel politics, the land they claim, and their tactics. Specifically, we propose that environmental governance behavior is motivated by the rebels' efforts to assert political authority, to sustain the land and livelihoods, and gain tactical advantages in their opposition against the state. In many ways rebel environmental governance is consistent with what we know about rebel governance more broadly. Nevertheless, we suggest it also features unique characteristics. Arguably, it is more fundamental than social service provision or the establishment of laws and courts: Environmental governance concerns the preservation of the very territory over which the rebels lay political claim; it reflects the rebels' recognition that the territory itself holds productive capacity, social and cultural meaning, and military value. Stripped of its natural worth, the territory loses its political capital, jeopardizing the entire rebel campaign. Faced with environmental threats, then, rebel groups may see environmental governance as essential to their survival and success.

We advance three interrelated arguments about the drivers of rebel environmental governance. These arguments jointly reflect our understanding that rebel groups are armed organizations whose ultimate aim is to gain political power within a delimited territory. Toward this end, rebels use both violent and nonviolent means to challenge the state, chipping away at the latter's military and political capabilities to eventually seize power (whether through regime overthrow or the assertion of their own statehood) or to induce concessions from

---

approaches that are more appropriate for the types of uncertainty that we face in managing climate risks can be embedded in governance. Assessments of governance barriers in Africa include elite capture (Kita, 2019), lack of local capacity in devolved governance structures (Musah-Surugu et al., 2019), and weak and unclear governance mandates (Shackleton et al., 2015) as key issues.

the government. Exercising political authority through local governance – including governance over their natural and built environment – is a major aspect of rebel strategy.

2.3.1 Rebels as Legitimacy-Seeking

First, *rebel environmental governance is a crucial part of rebels' contestation against the state, particularly as climate change heightens concern for environmental issues*. Rebel groups engage in environmental governance because they seek credibility as political authorities capable of managing complex challenges. Building a reputation as competent and credible authorities in turn enables them to garner legitimacy as political entities. The fight for political legitimacy is integral to rebel groups' political agendas, as numerous studies show (Cunningham et al., 2021; Duyvesteyn, 2017; Loyle et al., 2023, 2022; McWeeney et al., 2023; Podder, 2017; Terpstra, 2020).

Rebel groups are attuned to the various audiences that matter for their campaigns: the domestic populace, the state itself, and international stakeholders (Arjona, 2016; Cunningham & Sawyer, 2019; Duyvesteyn, 2017; Huang, 2016b; Mampilly, 2011; Van Baalen, 2021). Amidst deep domestic concerns about environmental degradation and climate-related threats ranging from drought, water access, and floods to deforestation and climate-induced migration across conflict-affected societies, rebels assert their governance authority over environmental issues to signal that they care about communities' welfare and are capable of taking action. If done successfully, rebel governance can also serve to highlight the incompetence or inaction of the state in the face of environmental threats, thus further boosting the rebels' assertions of authority vis-a-vis the state. Rebel groups also seek credibility in the eyes of state actors (Heger & Jung, 2017). For groups seeking or engaged in conflict negotiations, for example, environmental governance can cast rebels as actors that are capable of addressing national concerns and hence fit to assume state responsibilities. Finally, rebel groups seek international legitimacy and the external political backing that comes with it (Bob, 2005). External support is often a sine qua non for rebels' ultimate success (Salehyan et al., 2011), especially for secessionist organizations whose political outcomes hinge on international support for their sovereign statehood (Coggins, 2014; Fazal, 2011). In the age of climate change, rebel groups see environmental governance as an increasingly salient way to showcase their competence and legitimacy as eventual state authorities.

In all three domains – local, domestic, and international – there is a performative aspect to this particular logic of environmental governance. Rebels wish not only to engage in local environmental governance, but also to broadcast their environmental achievements to their audiences in a show of capability and authority. In this argument, environmental governance can be understood as an aspect of rebel image-making and public relations.

We thus expect rebel groups that display greater interest in domestic and international legitimation to be more likely to engage in environmental governance. Furthermore, such groups are likely to engage in environmental projects with the greatest visibility. Natural disaster management, whether climate-related or not, for instance, can get rebel groups instant recognition as governing authorities as disasters heighten domestic and international media attention for a time. The efforts of the Free Aceh Movement (GAM) in Indonesia following the 2004 Indian Ocean Tsunami, including coordinating with hundreds of aid organizations, catalyzed positive international attention for the group (Beardsley & McQuinn, 2009). The creation of institutions devoted to environmental governance is another visible endeavor, one requiring relatively little initial cost. For instance, a rebel "government ministry" dedicated to environmental affairs can exist on paper or on the Internet irrespective of any actual performance, helping lend bureaucratic authority to the rebel group. Similarly, rhetorical commitments can be announced with relatively little upfront cost, helping to cast the rebels as committed to local welfare and environmental protection – though failure to live up to commitments can be costly in the longer run if domestic and international audiences call out what they perceive as cheap talk.

Numerous rebel groups have taken action along these lines. Somaliland introduced a domestic law – the Environmental Management Law No. 79 (2018) – whose 81 articles aim at environmental governance, with the Ministry of Environmental and Rural Development overseeing implementation (Republic of Somaliland, 2018). Somaliland additionally boasts a Ministry of Environment and Climate Change, replete with a website showcasing its collaboration with foreign organizations such as World Vision and the Danish Refugee Council.[23] The National Movement for the Liberation of Azawad has made a commitment to respect international humanitarian law concerning the protection of the environment during military operations (De La Bourdonnaye, 2020, pp. 584–585). These examples highlight the important role that climate governance can play in rebel groups' efforts to garner legitimacy.

2.3.2 Rebels as Protectors of Land and People

Second, *rebel environmental governance reflects rebels' concern over land and livelihoods*. All rebel groups lay claim to specific territories and the communities

---

[23] https://moecc.govsomaliland.org/.

that reside in them. As such, rebel groups have a fundamental interest in ensuring that the territory itself preserves its value and that it benefits, rather than harms, the welfare of its residents. The preservation of the territory's natural environment, and concomitantly the protection of residents from environmental threats, is tantamount to defending the worth of the very territory for which they have mounted a fight against the state. The territory holds material, political, and symbolic capital, whether in its natural resource wealth and productive capacity, sustainable living spaces and livelihoods for its residents, or the cultural heritage that is tied to their land. Protection of the environment is also protection of the peoples' access to and enjoyment of the environment as a source of livelihood.

Rebel environmental governance is thus more territorially grounded, and arguably more fundamental to rebel politics, than is rebel social service provision or the creation of rebel administrative and judicial systems: It concerns the integrity of the territory's physical environment itself; the political and social life that unfolds within it is entirely contingent on the sustainability of the former. This is not an argument about rebels as environmentalists – whether they see themselves as such is immaterial to this argument; rather, rebels often see the primacy of environmental governance in their wider political project to gain authority over territories and their populations. It is in this light that we can understand reports of al-Shabaab's policy dubbed by observers as "eco-Jihadism": al-Shabaab in Somalia introduced a ban on single-use plastic bags due to the significant harm they cause to livestock in the pastoral areas where the rebel group operates (VOA News, 2018).

For many rebel groups, preserving the value of the land is also about enacting their historical and cultural claims to the land. Thus, Hamas's program to plant olive trees throughout Gaza may have had tactical benefits (described next), but perhaps equally important was the message of cultural resistance it imparted, as "these trees are a powerful symbol for Palestinian identity, with their roots representing ties to the land and their branches forced displacement from it" (Darwish, 2023). Similarly, Salween Peace Park, a forest conservation initiative launched by community leaders with the support of the Karen National Union (KNU), is both an environmental action and an effort to preserve the cultural heritage of their ancestral homelands in the face of military incursions by the Myanmar regime (Paul et al., 2023). Rebel environmental governance can thus derive from both material and nonmaterial drivers simultaneously.[24]

---

[24] Rebel groups can certainly govern without firm territorial control (Loyle et al., 2023). For instance, they may lay claim to a dam without controlling any territory or population beyond it, and use this control to manage water scarcity issues. Nevertheless, our argument returns to the basic idea that rebel organizations are ultimately fighting for state power, which is invariably tied to territory. To exercise governance authority over that territory is thus to govern over the land, its

This is not to say that rebels do not extract from or degrade the environment as well. For example, the Movement of Democratic Forces of Casamance (MFDC) regularly engages in environmentally destructive practices in Senegal, particularly in regard to illegal logging and timber trade in illicit markets (Medina et al., 2023). Yet, this group also engages in some conservation efforts. In recognition of their role in environmental degradation, particularly in the Fort Nord zone, MFDC leaders agreed to "an immediate stop to wood-cutting by combatants, interdiction on any other person cutting trees, surveillance of the whole zone in order that the measure be respected by everyone," as well as providing "information to and raising awareness of populations in line with this forest protection project" (quoted from MFDC meeting notes in Faye (2006, p. 56)). This behavior parallels our understanding of modern state governance, wherein governments across regime types manage their land and people while also extracting from them (Tilly, 1985).

Note that not all rebel groups claim to fight for the interests of a "people"; nevertheless, most rebel groups at least claim to represent the interests of some constituent population. We expect rebel groups that demonstrate concerns over social governance and civilian welfare to be more likely to engage in the management of territory via environmental governance, seeing it as central to supporting the livelihoods of the local population and enabling rebel politics more broadly. Conversely, rebel groups that demonstrate little regard for social governance and welfare, or those more prone to predatory behavior, are less likely to take up environmental governance.

2.3.3 Rebels as Fighters

Finally, *rebel environmental governance is about securing strategic and tactical advantages.* Such advantages can be more general, broadly affecting lives and livelihoods in the rebel-held territory, or more targeted, accruing specifically to the rebel group.

We have argued that environmental governance, including natural resource management and protection from climate-related environmental threats, can be crucial for communities' welfare. To the extent that rebel groups seek to assert authority over communities and demonstrate their ability to manage local issues, environmental governance is an important aspect of their broader

---

people, and its natural and built environment. More than land for land's sake, the argument underscores rebels' interest in sustaining the lives and livelihoods of the local residents over whom they govern; toward this end, they recognize the necessity of managing the environment.

political strategy. Effective rebel environmental governance undergirds rebel operations by sustaining both the rebels and local communities.

The advantages of rebel environmental governance go further, however, with direct tactical effects on rebel operations. Forest protection offers a clear example. Because forest cover shields rebel-held territories from aerial surveillance and other counterinsurgency operations, rebel efforts against deforestation can have direct tactical effects. An official of Hezbollah's reconstruction branch acknowledged this when he explained that trees and forests are military assets for the organization because they provide shelter and cover for their fighters. Hezbollah reportedly managed the planting of 7.3 million trees throughout Lebanon to restore the country's forests following their destruction in the 1975–1989 civil war (Karagiannis, 2015, pp. 185–186). The Free Aceh Movement (GAM) in Indonesia also used forests to evade the state military, leading to increased tensions and grievances with local forest communities who were subject to interrogations (McGregor, 2010, p. 24). Indeed, some observers note that rebel occupation of forested areas can lead to unintended forest protection as their operations make the forests inaccessible to state forces, ranchers, and logging companies (Munive & Stepputat, 2023).[25] On the other hand, rebel forest protection efforts vividly illustrate how environmental governance itself often becomes a site of armed politics. Counterinsurgents, well aware of the tactical value of forest cover, often directly target forests in their operations. The widespread use of chemical defoliants by U.S. forces in the Vietnam War is well known; the Indonesian government used deforestation as a counterinsurgency tactic in Sarawak and West Kalimantan (Peluso & Vandergeest, 2011) while Turkish forces have burned forests in the Kurdistan region of Turkey claimed by the PKK (Van Etten et al., 2008).[26]

Rebel groups vary in the extent to which they secure and control their own territory and create a "liberated" zone with their own governance and administrative system. If rebel groups engage in environmental governance for its tactical advantages, whether through forest protection or other measures, we can expect those groups with relatively firm territorial control to be more likely to engage in such measures. Rebel groups will be hard pressed to protect forests if the forests are not within their zones of control.

***

---

[25] For example, Mbade (2018) finds that during the Casamance conflict in Senegal, the rebel MFDC used forests as a safe haven, leading to an exodus of local communities and state agents from the area. This resulted in a growth of the forests during the conflict period.

[26] There are also contexts in which both rebels and government forces are accused of tactical deforestation, such as the dispute in Nagorno-Karabakh (Conflict and Environment Observatory, 2021).

These three arguments are interconnected and mutually reinforcing. Environmental governance aimed at projecting political authority also helps sustain the land and livelihoods and can enhance rebels' tactical defenses against counterinsurgents. Defending communities from state forces through reforestation is as much an environmental measure as it is a political and tactical one. While the theory parses the three mechanisms, in practice multiple motives jointly inform rebel groups' decision to engage in environmental governance and the particular governance measures they implement.

In many ways, rebel environmental governance mirrors state efforts: States, too, seek legitimacy and credibility both domestically and internationally, and they, too, have a general concern for preserving the physical integrity of territories within their jurisdictions. Both states and rebels are armed entities that must weigh the benefits of environmental protection against the allure of actions such as resource extraction, armament buildup and use, and industrial planning, all of which can harm the environment. Indeed, rebel governance scholarship elucidates how rebel motivations are both parochial, enacted in response to local needs and contingencies, and also more general, spurred by a fundamental desire to *be* like a state in order to replace the state; the idea of mimicry of the state as part of a revolt against the state has an established place in the rebel governance literature (Arjona et al., 2015; Florea, 2020; Huang, 2016b; Mampilly, 2015). To what extent rebel groups pursue environmental governance in direct response to, or as a pushback against, state environmental governance is an important question for future research, as we discuss in the Element's conclusion. Nevertheless, rebel groups also have a distinct set of political interests, as we have asserted earlier: The need to gain local and international legitimacy despite their "non-state" status, to ensure land is preserved for a post-conflict future, and to leverage the environment for tactical gains in the present, all in the context of what is typically a severe resource disadvantage vis-à-vis their state adversaries.

2.4 Empirical Expectations

Our approach yields three broad propositions that mirror the foregoing discussion.

- Rebels that seek greater legitimacy (with the local population, with the state, or with the international community) are more likely to engage in environmental governance.
- Rebels that are more invested in the livelihoods of the local community are more likely to engage in environmental governance.

- Rebels that can gain strategic or tactical advantage by utilizing features of the natural environment are more likely to engage in environmental governance.

In the next section, we introduce the Rebel Environmental Governance (REG+) data, providing a first look at patterns of environmental governance by rebel groups.

3 Introduction to the REG+ Data

3.1 Sample and Scope

The Rebel Environmental Governance Dataset (REG+) draws from the widely used Uppsala Conflict Data Program Armed Conflict Dataset (version 23.1). The data include all instances of civil war worldwide that were ongoing in or started after 1989, generated twenty-five battle deaths in any year, and have an identified rebel group and state fighting over territory or for independence (secessionist groups) (Davies et al., 2023; N. P. Gleditsch et al., 2002); the data excludes wars fought for control of the central government.[27] Our unit of analysis is a state-rebel group dyad. For each armed actor, we code if the group engaged in each environmental governance behavior of interest during active periods of conflict.[28] Following most quantitative civil war studies, conflict periods include breaks in fighting (dropping below the battle-death threshold) for three years or less. Breaks lasting four years or longer are noted but will not be coded in the data.[29] Evidence of behavior during these breaks or after the conflict has ended is not included in variables.

The dataset includes 162 rebel groups across four geographic regions.[30] The 162 cases show significant variation in the regions they cover, with only the Americas not represented in the data. The omission of the Americas is a result of the absence of civil war over territorial control in the region during this period as defined by UCDP. While conflicts in Asia represent the majority of cases in the

---

[27] This limitation to "territorial" conflicts is due to resource constraints in what turned out to be a highly intensive data collection exercise. We selected these disputes, as opposed to center-seeking conflicts, because these wars are likely to feature rebel groups making claims to specific territory over which they may see benefits of establishing governance distinct from established authority (i.e. they are not trying to take over existing governance structures per se).

[28] In a number of cases, we find evidence of the group engaging in REG+ in times outside of the "active conflict" period (years in which there were 25+ battle deaths). In those cases, we attempt to verify that any particular behavior occurred during the active conflict period. We explore REG+ outside the active conflict periods more in the vignettes in Section 4.2.

[29] Conflicts with breaks are still coded as a single rebel group; we attempt to include only rebel behavior that occurs in the active periods.

[30] UCDP includes two groups (LPR and DPR in Ukraine) in two separate disputes with different dyad ids in 2014 and 2015. We consider these LPR in both disputes as the same actor, and DPR in both disputes as the same actor.

data, Africa, the Middle East, and Europe also constitute a large portion of the cases. Online Appendix Figure C1 shows the geographic variation of rebel groups in the sample based on the region classification by Gleditsch and Ward (1999).

The rebel group-state dyad cases vary widely on a number of dimensions relevant for our understanding of civil war dynamics. The cases covered in our data feature different types of conflicts, including both small and large civil wars, although the majority of conflicts are lower-intensity wars (Online Appendix Figure C2).[31] On average, the cases in our data generated 2,590 battle-related deaths. Most wars were continuous events, however, thirty-four rebel groups engaged in recurrent conflict, where there were multiple-year breaks in fighting.[32] We also see wide variation in conflict duration in our sample (Online Appendix Figure C3), with the shortest wars lasting for one year, and the longest for more than four decades. On average, rebels in our sample fought for about six years.

3.2 REG+ Variables

As previously noted, we adopt a broad and inclusive definition of rebel environmental governance. Our operationalization and coding of REG+ mirrors this conceptualization, enabling scholars to identify and theorize a wide range of rebel behavior. A coding narrative is provided for each rebel group in the dataset with references, which allows scholars to narrow the scope of the definition of environmental governance as needed for specific research questions. This is especially important for scholars who seek to employ the REG+ data to further understand how rebel behavior may impact climate mitigation efforts (e.g. through the effects of conservation) or on adaptation efforts where rebel actors enact agricultural policy or manage access to the resources needed to implement adaptation programs.

In identifying the relevant behaviors that constitute rebel environmental governance, we take as our starting point the work of the Intergovernmental Panel on Climate Change (IPCC). According to these initiatives there is a wide set of policies, activities, and responses that comprise environmental

---

[31] The majority of cases (72 percent) are rebel groups involved in minor armed conflicts, which produced between 25 and 999 battle deaths total. In 22 percent of our cases, rebel groups participated in larger wars (i.e. those that led to between 1,000 and 9,999 battle deaths total). Only 6 percent of the conflicts in our data did rebel groups fight in wars that led to at least 10,000 deaths due to battles.

[32] On average, rebel groups were active in a conflict for about six years, though some only fought for one year (such as the Anjouan MPA in Cameroon), and others for decades (such as the Karen National Union in Myanmar). Despite our coding including conflicts starting or ongoing after 1989, there is a wide coverage across years with some conflicts in our sample starting as early as 1961.

governance, including climate action, disaster risk reduction (DRR) and pursuit of the Sustainable Development Goals (SDGs).[33] Climate action has been typically understood to include climate mitigation and climate adaptation activities with new provisions related to loss and damage (L&D) being increasingly articulated and funded (Broberg, 2020; Hossain et al., 2021).

To catalog rebel behavior in environmental governance, we identify four broad types of actions. The first set of variables captures rebel management of the natural environment – the focus of many adaptation efforts – including actions relating to agricultural and land practices, food security, water security, and conservation. Second, to capture rebel responses to acute crises related to natural and climate hazards, we collect information on rebel provision of migration assistance to climate-affected populations, as well as disaster management efforts (post-disaster). Third, we capture aspects of strategic framing and institution-building by coding for rebel use of environmental and climate change rhetoric, as well as the creation of rebel political offices or bureaucracies charged with environmental or disaster risk management. Fourth, as a way to begin to capture the broader political dynamics of rebel environmental governance, we collect data on rebel cooperation with other actors in environmental governance, including states, NGOs, and IGOs. In this way we capture a wide range of rebel environmental governance behaviors, including those with spillover effects for climate action and those pursued in response to climate change.

Specifically, we code the following environmental governance behavior by rebel groups:

- *Agricultural practice and land management*: This variable captures whether the rebel group provided governance related to agricultural practices and/or land management (such as land tenure management and enforcing specific crop planting practices).

  > Example: The Syrian Democratic Forces implemented agricultural reforms in northern Syria to promote sustainable agriculture in regions with water insecurity and a high risk of drought. The group embraced small-scale farming and worked to prevent monoculture and water-intensive agriculture. It has also banned chemical fertilizers and distributed organic fertilizers with the explicit aim of restoring soil and water quality.

---

[33] The SDGs are a set of goals adopted by UN members as "a universal call to action to end poverty, protect the planet and improve the lives and prospects of everyone, everywhere." www.un.org/en/exhibits/page/sdgs-17-goals-transform-world.

- *Water access and security*: This variable captures whether the rebel group provided governance related to water access and water security.

    Example: The Moro Islamic Liberation Front in the Philippines cooperated with communities in the Ligawasan marshes to enforce traditional indigenous laws regarding fishing, farming, and other uses of water. These practices included banning electric and chemical fishing and fly catching and enforcing open access to the marshlands for all communities.

- *Food access and security*: This variable captures whether the rebel group provided governance related to food access and security.

    Example: Al-Itihaad Al-Islamiya in Somalia engaged in agricultural enterprises and management of fishing rights in coastal areas.

- *Conservation*: This variable captures whether the rebel group provided governance related to environmental and/or wildlife conservation.

    Example: The Karen National Union in Myanmar created nature reserves in most of the districts under their control. They also supported and provided certifications for a series of community-driven conservation efforts in their regions of control.

- *Migration assistance related to environmental hazard*: This variable captures whether the rebel group provided governance related to human migration caused, at least partially, by an environmental hazard (such as drought, earthquake, flood, desertification, soil degradation).[34]

    Example: Polisario in Morocco managed refugee camps in Algeria that house a majority of the Sahrawi population from West Sahara not under Moroccan control. Increasing drought and desertification–combined with the border wall created by Morocco–have made the region all but uninhabitable.

- *Disaster risk management*: This variable captures whether there were rebel efforts at natural and climate-related disaster management, including the provision of disaster relief services.[35]

---

[34] We include responses to some hazards where the links to climate change are less direct or well understood, specifically earthquakes. At least indirectly, earthquakes may be influenced by climate change processes such as droughts (Argus et al., 2017) and the potentially associated changes in groundwater extraction (Amos et al., 2014).

[35] We have included all cases where there was disaster relief from a natural hazard regardless of the attribution to climate change. For example, in the case of tropical cyclones, only some characteristics of the hazards are associated with climate change, namely increased rainfall, and where the flooding is compounded by sea level rise (Knutson et al., 2021). We also include cases like the one highlighted here where we document the observed behavior after a tsunami. Tsunamis are not directly the result of climate change; however, climate change can indirectly increase the risks. For example, there is some evidence that sea level rise can expand the damage zone although the added impact depends on the magnitude of the tsunami (Dura et al., 2021).

Example: The Free Aceh Movement (GAM) in Indonesia responded to the devastating tsunami that hit Indonesia in 2004 by providing initial support and access to international humanitarian aid in the regions under its control.

- *Rhetoric*: This variable captures whether the rebel group used rhetoric related to climate change or environmental issues to justify the conflict.

  Example: The Free Papua Movement (OPM) in Indonesia operates as a government in exile which emphasizes climate issues and environmental governance as a central component of its campaign to gain statehood. A political component of the OPM based in Vanuatu, the United Liberation Movement for West Papua, announced at the 2021 Glasgow COP26 conference the "Green State Vision" for West Papua.

- *Disaster Management Institution:* This variable captures whether the rebel group created a governance body charged with disaster management.

  Example: The Fatah government, fighting against Israel, includes the Environment Quality Authority (EQA) which was charged with disaster response, among other functions.

- *Environmental Institution:* This variable captures whether the rebel group created a governance body charged with managing environmental concerns.

  Example: The Donetsk People's Republic in Ukraine created a number of institutions including a Committee of Land Resources and Ministry of Agriculture and Food Production.

We further code information on the political dynamics of rebel cooperation related to environmental governance:

- *Cooperation*: This set of variables captures whether the rebel group cooperated with another actor on environmental issues. We code a separate indicator for cooperation with each of the following:

  - domestic NGO,
  - international NGO,
  - international organization,
  - religious organization,
  - domestic state government (at any level),
  - foreign states.

    Examples: The UFLA in India worked with Jatiya Unnayan Parishad (JUP), a local development organization, to enable service provision. The Free Papua Movement (OPM) in Indonesia cooperates with several types of actors, including international NGOs, such as the International Parliamentarians for West Papua (IPWP), local churches, the Indonesian government, and the United Nations.

3.3 Sources

To collect the data, we rely on publicly available sources retrieved using Google Scholar, the University of Maryland Library System, and Google Search. Coders are given search procedures outlining the minimal search requirements for any case variable. For each variable, coders are also provided with specific search terms to facilitate and better standardize each coding. Refer to Online Appendix A for full details on the coding procedures.

All sources were documented and ranked based on a tier system of source types. Specifically, our high-ranking sources (tier 1) are those gathered from peer-reviewed journal articles and books, Ph.D. dissertation or master's theses, policy reports from governments, IGOs, IOs, and NGOs, and news reports. Our low-ranking sources (tier 2) are those gathered from non-peer-reviewed books and articles, and other publications such as websites and blogs. The majority of the information comes from academic sources and policy reports. Online Appendix Figure B1 shows the distribution of sources used to code all positive cases of each variable across all source types.

We further assess the quality of the sources used by each variable of interest. Overall, the majority of our positive codings come from high-quality sources (tier 1), with only disaster risk management having over 23 percent of the codings depend on lower-quality sources (tier 2). Data quality remains high across different regions and time periods, suggesting no one case or set of cases drives our use of lower-quality sources. Online Appendix Figures B2 to B4 show the breakdown of source type by variable, time period, and region, respectively. In short, this information suggests that the source quality for the data is on average high.

3.4 Patterns of Rebel Environmental Governance

Our novel data suggests substantial variation in rebel governance over environmental and climate concerns. Figure 2 shows global variation in the number of rebel groups engaging in REG+ behavior. Online Appendix Figures C4 to C6 provide similar maps for rebel environmental institutions, rhetoric, and cooperation. Note that the lack of cases in Latin America is due to the fact that UCDP does not identify any rebels as fighting over "territory" in this area during the period under study. Future iterations of the REG+ will include rebel groups that fight for control of the central government, which will capture a number of insurgencies in Latin America, many of which do engage in REG+.

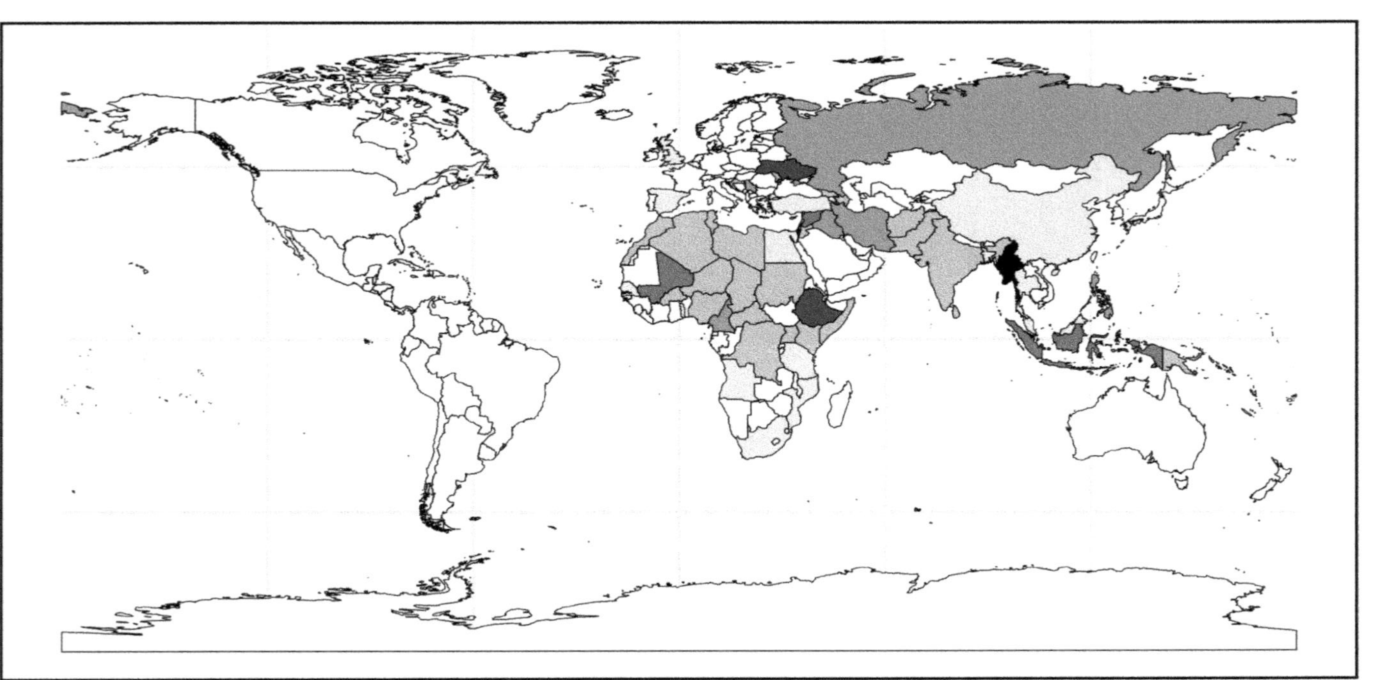

Figure 2 Geographic spread of rebels employing REG+ behaviors

[**Note:** White=no rebel groups with territorial claims]

We find a significant distribution of environmental governance behaviors across our cases, suggesting rebel environmental governance is now widespread. Nearly half of the rebel groups in our sample (43 percent of groups) engage in at least one behavior related to environmental governance (including water security, food security, conservation, migration assistance, agricultural management, and disaster response). This figure is higher if we include rebel groups that built institutions charged with environmental affairs or disaster management, and those that used environmental rhetoric. Some groups, such as the SDF in Syria (and its predecessor, the PYD), engage in all of these behaviors. Figure 3 provides a histogram of groups by the number of distinct REG+ behaviors they use. The majority of groups use more than one behavior, and a significant proportion engage in four or more behaviors.

We see variation across the unique types of environmental governance. Figure 4 shows the frequency of the REG+ behaviors in the top panel. In the middle panel, we show the distribution of values for climate and environmental rhetoric and the institution variables; the lower panel shows the cooperation variables.

The most common form of REG+ is agricultural management, with well over a third of the rebel groups in our sample engaging in this behavior, followed by food and water security at about 32% and 26%, respectively. These findings suggest rebels are involved in some of the most fundamental aspects of civilian welfare provision in the face of climate and environmental hazards. Most of these activities occur relatively evenly across regions, although governance

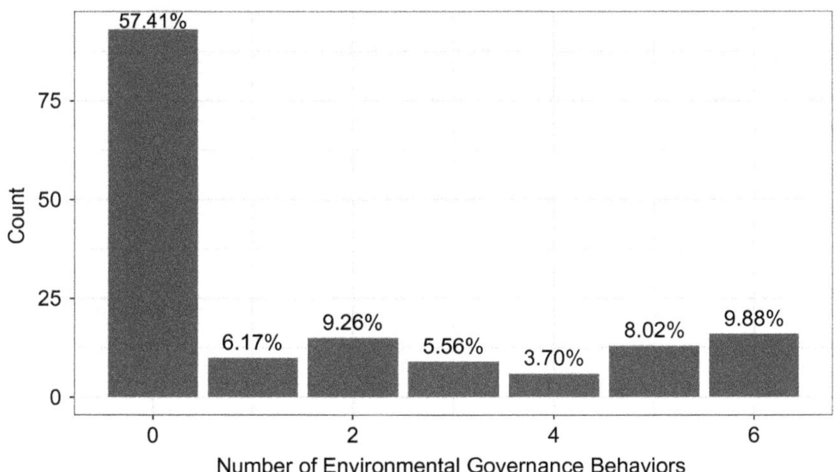

Figure 3 Distribution of groups by the number of REG+ behaviors used

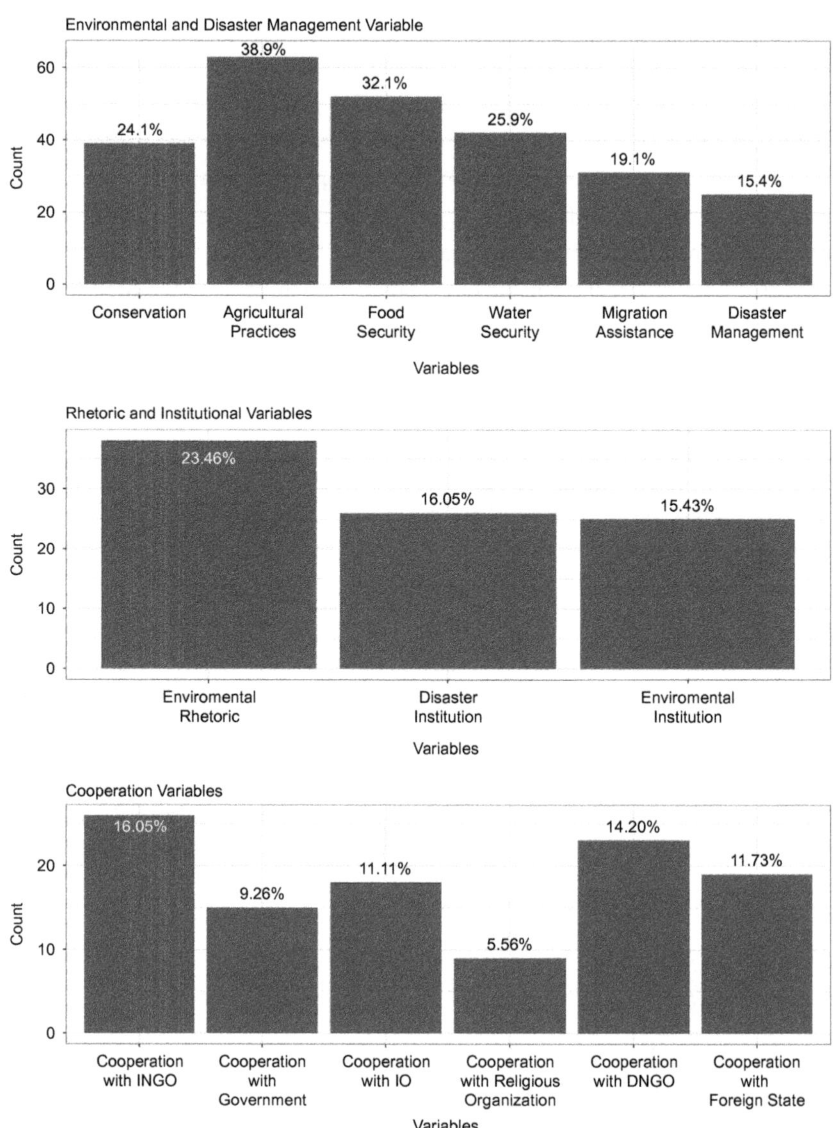

Figure 4 Rebel Environmental Governance variables by frequency

related to food security is more common in Africa, with twenty-one out of fifty-two cases of this occurring there.

A significant proportion of rebel groups, nearly one in four, engages in conservation in the midst of conflict. This is evenly distributed globally. Many conservation efforts focus on the protection of forests, for example by the Republic of Artsakh (Nagorno Karabakh) which fought against Azerbaijan.

However, the Karen National Union in Myanmar also works with local NGOs to support wildlife and biodiversity conservation (KESAN, n.d.; Pearce, 2020). Of the thirty-nine rebel groups that engage in conservation, about 26 percent (ten groups) also engage in deforestation at some point.[36] This complex relationship between rebels, conservation, and environmental degradation and destruction is highlighted in the qualitative vignettes in Section 4.

Over one in five rebel groups have used some type of environmental or climate-related rhetoric. This ranges from claims of government abuse of the environment, such as those made by the Eastern Turkistan Islamic Movement against China (Reed & Raschke, 2010), to advancing a forward-looking narrative, such as the statement by the General Secretary of Hamas in 2010 that "today humanity is confronting this great and serious climatic threat" (Karagiannis, 2015, p. 186).

A number of rebel groups also invest, at least on some level, in the creation of institutions related to environmental and/or disaster management. Some of these are clearly well established and functioning institutions, such as the PUK controlled Kurdish Ministry of Agriculture and Irrigation in Iraqi Kurdistan, while others appear to be more aspirational. Yet even institutions that may not be functional in practice can reflect meaningful signals and investment on the part of the rebel group.

Rebel cooperation with other actors is fairly common – about 25% engaged in cooperation with at least one of the other types of actors we consider. Perhaps surprisingly, rebel groups cooperate most often with international NGOs, at 16% of the cases, followed by domestic NGOs at about 14%. The relative frequency of cooperation with international NGOs reflects the significant role played by such organizations in climate adaptation in conflict contexts. International NGOs may be more agile in working with rebel groups than are foreign governments and IOs, which may face greater legal and bureaucratic impediments. Rebel groups even cooperate with the state adversary in the conflict – we see this in about 9% of our cases. These observations collectively attest to the prevalence of hybrid governance – one that includes rebel groups – in responding to environmental issues in conflict contexts.

Our goal in creating the REG+ was to reflect the wide range of actors that play a role in addressing climate-related challenges in conflict-affected areas. We do so by specifically integrating rebel groups into the broader scholarship of multilayered governance around environmental issues. The REG+ demonstrates that rebel groups engage in a variety of different types of environmental

---

[36] The REG+ data also includes an indicator on deforestation behavior, but not broader instances of environmental degradation.

governance across contexts. Thinking about rebel groups as active agents in environmental governance opens new avenues for scholarly research and policy interventions.

4 Exploring the Determinants of Rebel Environmental Governance

In Section 3, we showed that rebel groups engage in a variety of environmental governance activities. This section will examine which groups engage in environmental governance, with a focus on the three arguments presented in our framework.

We present two types of analysis – quantitative and qualitative. First, we use the REG+ data to test the first and second general implications drawn from the framework we advance in Section 2 (focused on rebels' desire for legitimacy and rebels' investment in their land and people). The quantitative analysis centers on explaining the occurrence of environmental governance. Second, we provide two case vignettes for each of the three arguments, which highlight the connections between these dynamics and rebel provision of environmental governance. While we are not able to directly test the third implication with existing quantitative data, our vignettes demonstrate support for our theoretical expectations by illustrating these dynamics at work in the cases. We do not suggest that our analyses are definitive tests of our implications, but instead an initial exploration that leverages novel data. We find the results are supportive of our broad empirical expectations. We offer paths toward a clear research agenda at the intersection of climate and the environment and rebel governance (discussed further in Section 5).

4.1 Quantitative Analysis Using REG+

There is no prior quantitative literature that has established a baseline model of factors likely to impact environmental governance by rebels. As a first step, we examined the distribution of rebel groups that engage in environmental governance among commonly studied conflict and governance factors.[37] Our dependent variable is the use of any of the REG+ behavior variables (which include water security, food security, conservation, migration assistance, agricultural management, and disaster response). We examine the occurrence of any

---

[37] These include duration, intensity (battle deaths), the number of rebel actors in the conflict, external support to rebels, regime type/democracy, and one-sided violence by rebels and states. The number of rebel groups is negatively associated with the provision of REG+, but this finding is not robust to the inclusion of our main independent variables in this section. We include a second model in Online Appendix Table D1 that excludes external support as this variable has more limited data coverage than the others; findings are similar.

environmental governance for two reasons. First, our theoretical framework does not necessarily yield unique predictions for the different types of environmental governance. Second, the opportunity for rebels to engage in any specific type of behavior is conditioned by their physical environment and the hazards they are facing. Examining this broadly with a measure of any environmental governance allows that rebel groups may have different opportunities and constraints on their provision of certain types or the number of behaviors that are unrelated to rebel motivations. Utilizing the UCDP data, which covers our full range of cases, we find only that conflict duration is statistically associated with the provision of REG+ at conventional levels (0.05 significance) (Online Appendix Table D1). We include conflict duration in the subsequent models.[38]

4.1.1 Rebels Seeking Legitimacy

Our first empirical expectation is that rebel groups that are legitimacy-seeking are more likely to engage in REG+. A number of scholars have advanced legitimacy-seeking as central to understanding the behavior and structure of rebel groups.[39] Of particular relevance, many of these studies have found that rebel groups seek to engage with the international community as representatives of their constituent groups and as rightful "governors" of their claimed territories and populations. While rebel diplomacy most directly reflects rebels' pursuit of international legitimacy, rebels also conduct diplomacy for domestic consumption and legitimation (Arjona et al., 2015; Duyvesteyn, 2017; Florea & Malejacq, 2023; Furlan, 2020; Heger & Jung, 2017; Huang, 2016a; Loyle et al., 2023, 2022; Mampilly, 2011; Staniland, 2012; Stewart, 2021; Worrall, 2018).

The REG+ data allows us to explore the expectation that rebel groups use environmental governance as part of their efforts to establish their credentials as legitimate governors. Our first analysis uses the Rebel Quasi-state Institutions (QSI) Dataset (Albert, 2022), which offers several indicators related to rebel legitimacy-seeking behavior. The temporal scope of QSI is more limited than REG+ (ending in 2012), and thus we have a fairly limited sample of overlapping cases (58 total), which should be considered in interpreting results.

We combine several variables to create a measure of rebel groups seeking international legitimacy. The indicator is a dichotomous variable coded as "1" if

---

[38] In a series of bivariate models using the standard conflict factors (Online Appendix Table D2), we see a small, positive, and statistically significant coefficient of conflict intensity, and negative and statistically significant effect of the number of other rebel groups in the conflict and a liberal democracy index.

[39] These studies focus on a number of outcomes, as well as on the processes through which rebels work to develop legitimacy (Cunningham et al., 2021; Duyvesteyn, 2017; Huang, 2016b; Jo, 2015; Loyle et al., 2022; McWeeney et al., 2023; Podder, 2017; Terpstra, 2020; Terpstra & Frerks, 2017).

Table 1 Logistic regression of legitimacy-seeking behavior
on environmental governance

	Dependent variable:
	Environmental governance
Conflict duration	0.111^{**}
	(0.043)
Legitimacy-seeking behavior (any)	1.745^{***}
	(0.649)
Constant	-2.321^{***}
	(0.642)
Observations	58
Log Likelihood	-29.131
Akaike Inf. Crit.	64.261

Note: $^{*}p < 0.1$; $^{**}p < 0.05$; $^{***}p < 0.01$

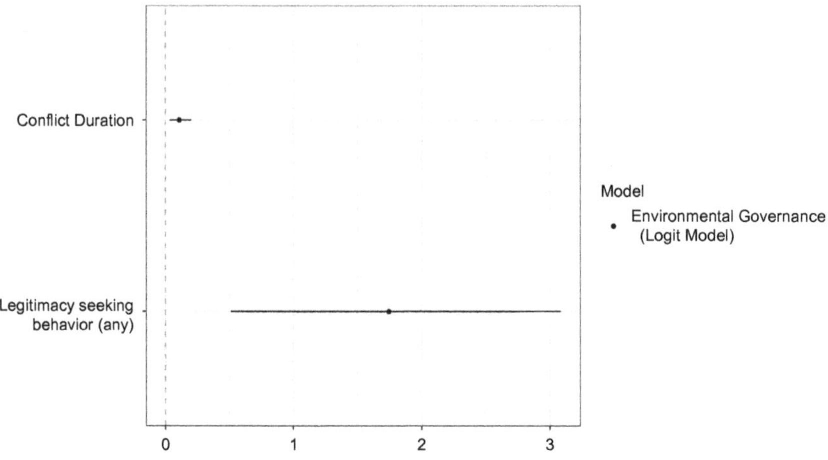

Figure 5 Coefficient plot for Table 1

the group has an embassy, joined an international organization, and/or attempted to join an international organization at any point that the rebel group was active. For example, Polisario in Morocco operates an embassy in Mexico and the Ogaden National Liberation Front (Ethiopia) and the Free Papua Movement (Indonesia) both belong to the Unrepresented Nations and Peoples Organization. All three of these organizations are coded as legitimacy-seeking by this measure. Just over 40 percent of rebel groups in the combined sample (twenty-three groups) engage in some diplomatic effort. In our framework, we link both domestic and international legitimacy-seeking to environmental governance. Empirically, we examine domestic legitimacy-seeking as part of our probe of the second argument on investing in the community, as discussed in the next section.

Using logistic regression, we examine the probability of a rebel group providing environmental governance (water security, food security, conservation, migration assistance, agricultural management, or disaster response), including whether the group engaged in a legitimacy-seeking activity and the duration of the conflict (in years). Table 1 shows the results of this analysis and Figure 5 shows the coefficient plots.[40]

We find a positive and statistically significant relationship between legitimacy-seeking activities and rebel environmental governance, as well as between conflict duration and REG+. Rebel groups that work to build international legitimacy are about 29% more likely to also engage in governance of the environment than groups that do not.

4.1.2 Rebels Investing in the Local Community

Our second empirical expectation is that rebels that are investing in ensuring the livelihoods of the local community will also have incentives to protect and care for the land to which they lay claim. We again utilize the QSI data (Albert, 2022) to capture two dimensions of investment in the local population. The first indicator is related to the provision of goods and services. This dichotomous variable measures whether the group provided education, health, housing, aid, and/or infrastructure services. A variety of groups provide these services, including Hamas (health), the RCSS in Myanmar (housing), PMR in Moldova (aid), and the PUK in Iraq (infrastructure services). Just over 40 percent of rebel groups in the overlapping sample provide goods and services (24 of 58). The second dichotomous indicator captures the provision of social governance, which includes whether the group had a "government," had a constitution, was organized like a state government, or provided judicial services. Again, we see a variety of rebel groups with these characteristics, including the PIRA in the United Kingdom (constitution), the KDP in Iraq (government-like), and the KIO in Myanmar (judicial services). About 62 percent of the groups in the overlapping sample (36 of 58) engaged in some form of social governance. While we use these measures primarily to examine the

---

[40] Results are similar if we use linear probability modeling (Online Appendix Table D3) and with a probability model of legitimacy-seeking behavior on a count of the environmental governance variable (Online Appendix Table D4).

Table 2 Logistic regressions of population investment behaviors on environmental governance

	Dependent variable:	
	Environmental governance	
	(1)	(2)
Conflict duration	0.111***	0.098**
	(0.042)	(0.041)
Social governance behavior (any)	1.494**	
	(0.714)	
Goods and services provision (any)		1.349**
		(0.631)
Constant	−2.576***	−2.028***
	(0.765)	(0.570)
Observations	58	58
Log Likelihood	−30.560	−30.700
Akaike Inf. Crit.	67.119	67.400

Note: $*p < 0.1$; $**p < 0.05$; $***p < 0.01$

argument about rebels investing in the local community, they also roughly capture rebel groups seeking domestic legitimacy, as the two are closely related: Part of the impetus for rebel governance and social service provision is to establish their credibility locally and garner popular legitimacy.

Employing logistic regression for each of these two factors, controlling for the duration of the conflict,[41] we find that both goods provision and social governance are associated with a higher probability that a rebel group will engage in environmental governance. Table 2 provides this analysis and Figure 6 shows the coefficient plot.

Rebel groups that establish these governance structures are 26% more likely to also provide environmental governance, while those that offer goods and services are 24% more likely to provide environmental governance.[42] Figure 7

---

[41] We model each variable separately due to the relatively high correlation between the two independent variables and small sample size, which creates concerns about multicollinearity when using both measures in the same model. In our combined sample, the correlation between these two variables is 0.44. In the Online Appendix Table D5 we run a regression with both variables. The coefficients remain constant, however due to large standard errors they are only significant at $p < 0.17$.

[42] Results are similar with linear probability modeling for the dichotomous measure of REG+ (Online Appendix Table D6) and with a linear probability model on a count of REG+ behavior

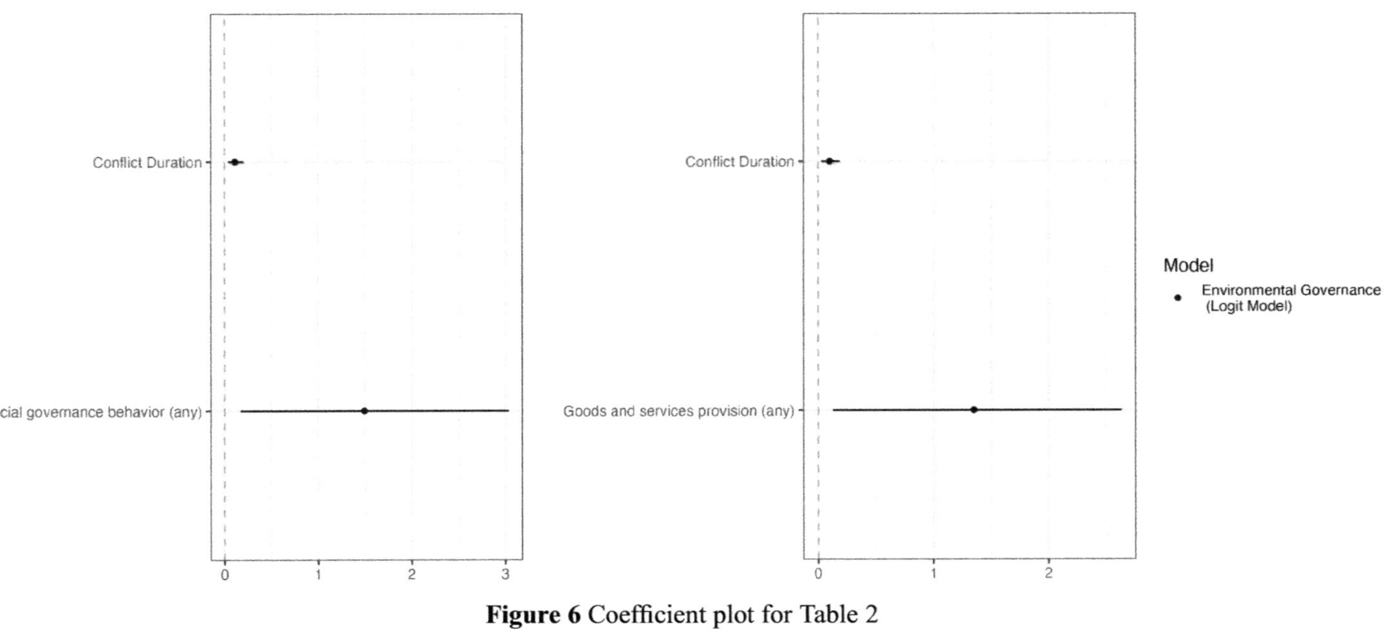

Figure 6 Coefficient plot for Table 2

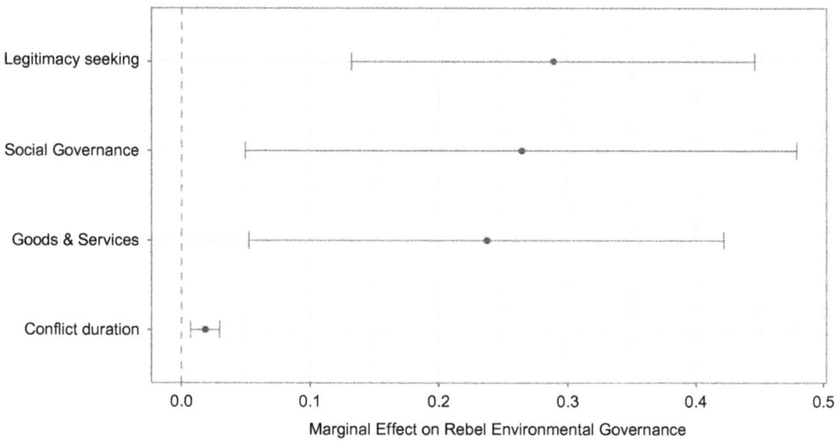

Figure 7 Average marginal effects from Tables 1 and 2

shows a comparison of the average marginal effects of these factors for both models.

The legitimacy-seeking, goods and services, and social governance factors have similarly sized effects on the probability of rebels engaging in environmental governance. The marginal effect of conflict duration shows the effect of a one-year change (based on Table 1). If we look at the change in duration from the first to third quartile of the data (one year to seven years), the effect is about 18%. This suggests that while longer conflicts are associated with a greater chance of rebels providing environmental governance, the legitimacy-seeking, goods and services, and social governance factors have a larger impact.

4.1.3 Rebels Seeking Tactical/Strategic Advantage

Our statistical analysis does not speak directly to the argument that rebels use REG+ for strategic or tactical advantage. We can make a very tentative link to rebel capacity, which may relate to strategic and tactical needs. The findings on conflict duration may suggest that rebels with the capacity to fight longer may also have the capacity to engage in more REG+ activities. Yet without further

---

(Online Appendix Table D7). Using the Foundations of Rebel Emergence (FORGE) (Braithwaite & Cunningham, 2020) dataset, we also examine how organizational origins, ideology, and goals of rebel groups may impact their propensity to provide environmental governance. The FORGE sample has a better overlap with the REG+ cases, yielding a sample of ninety-five rebel groups. We do not find a significant relationship between rebel groups that start from non-violent (more representative) organizations (such as nonviolent movements and youth organizations), nor those that pursue goals related to the rights of a constituent population, independence or a specific ideology (Online Appendix Table D8).

analysis, it is not clear how exactly such capacity will intersect with strategic needs of rebels.

4.1.4 Summary and Interpretation of Quantitative Findings

Our primary goals in creating the REG+ dataset is to document the use of environmental governance by rebels and open paths for further inquiry about these dynamics and their broader political impacts. We have suggested three explanations for rebel environmental governance – legitimacy-seeking, investment in the land and local community, and strategic/tactical advantage. We find initial quantitative support for our argument about legitimacy-seeking and investing in the local population serving as viable explanations for rebel environmental governance. Groups that seek domestic and international legitimacy, as well as those that develop local governance structures, are most likely to engage in environmental governance. It is important to note that this research design does not allow us to speak to causality; instead, we are able to establish basic empirical patterns. Nevertheless, our case vignettes of six rebel organizations, offered below, provide more detailed analysis that help establish the plausibility of our three-part framework.

4.2 Qualitative Vignettes on REG+

In addition to the broader trends in rebel environmental governance, we explore here a set of case vignettes that illustrate the key dynamics we advance on why rebels govern over the environment. Our cases are selected based on information gathered while coding REG+. Our primary goal in presenting these vignettes is to illustrate the rich variety of rebel environmental experiences (as opposed to hypothesis testing). We identified cases with environmental governance which included our dynamic of interest, and then explored primary and secondary material to document the causal pathways driving each dynamic.[43] This is a convenience sample that is selected to illustrate the presence of our theorized mechanisms and to generate additional empirically testable hypotheses. Whereas our data enable us to document that REG+ activity is widespread across contemporary armed conflicts and to establish basic patterns across cases, our case vignettes offer a more contextualized, detailed, and nuanced analysis. In most cases of rebel environmental govern-ance, we can identify an interplay of potential drivers. These cases also allow

---

[43] It may also be instructive for future research to examine cases where we observe the use of rebel governance in other domains but not the environment (such as by the RCSS in Myanmar) or the reverse (such as the UFLA in India). The REG+ case sheets include details on cases both with and without evidence of environmental governance that can facilitate such comparisons.

us to examine the dynamics of rebel governance related to climate and the environment during times when rebels are not necessarily considered "active" in the armed conflict data, which is based on a battle-death threshold. For each case we identify the "active" years, but also explore environmental governance dynamics around this period. We organize the vignettes around our three main arguments: legitimacy-seeking, investment in the local community, and tactical advantage.

4.2.1 Rebels Seeking Legitimacy

In the cases of Polisario and the Republic of Artsakh, we illustrate how rebel groups engage in environmental and climate governance as a way to garner domestic and international legitimacy.

CASE 1: POLISARIO (WESTERN SAHARA/MOROCCO)

Frente Polisario (Polisario Front) has been fighting for the independence of Western Sahara since Spain's withdrawal from its former colony and Morocco's subsequent annexation of the territory in 1975. Announcing an armed struggle for self-determination, Polisario proclaimed the Sahrawi Arab Democratic Republic (SADR) in 1976. UCDP codes the Polisario as active in conflict (i.e. generating twenty-five battle deaths or more in a year) in 1975 to 1989. While armed confrontations are now rare, Polisario continues to challenge Morocco's claims to Western Sahara through political and diplomatic channels up to the present.

In practice, Morocco controls the majority of the territory of Western Sahara while Polisario controls its sparsely-populated "liberated zone." Since the 1980s Morocco has effectively physically separated the Moroccan-controlled west from the Polisario-controlled east through the construction of a 2,700-kilometer fortification, known as "the Berm." With a large number of Sahrawis having fled the territory and settled in refugee camps near Tindouf, Algeria, Polisario administers the refugee camps and their Sahrawi residents with its own political system and social services as a government-in-exile.[44] Figure 8 shows the Berm in Morocco.

---

[44] For a backgrounder on the Western Sahara conflict, particularly regarding the issue of natural resources and international law see Kingsbury (2018).

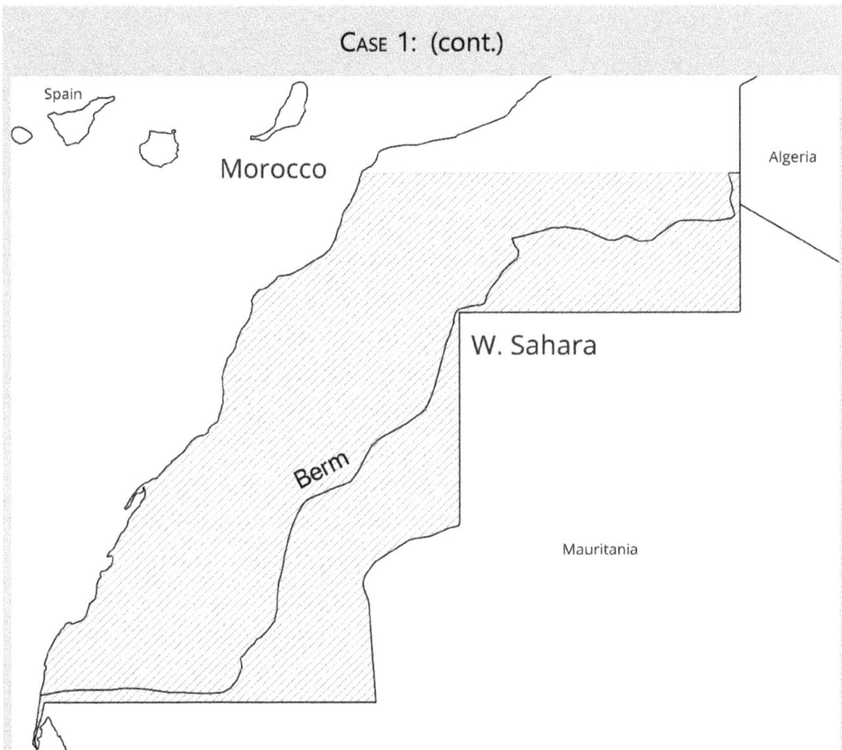

Figure 8 Western Sahara and Morocco with the Berm
(black line in shaded area)

Polisario's mission to gain external legitimacy, ultimately via a formal international recognition of the SADR, is explicit. The organization invokes a 1975 International Court of Justice (ICJ) decision rejecting Morocco's and Mauritania's claim to Western Sahara, as well as a 1979 UN General Assembly resolution recognizing Polisario as the representative of the Sahrawi people, as providing a legal basis for its sovereignty claims over Western Sahara (Allen & Trinidad, 2024). Demonstrating its deeply contested status, the SADR enjoys bilateral recognition by several dozen states and has been a member of the Organization of African Unity (OAU), now the African Union, since the 1980s, but has not been recognized by any permanent member of the UN Security Council.

The confluence of armed conflict, natural resource wealth, and climate change has had a devastating impact on Western Sahara and its residents' livelihoods (Porges, 2020). Polisario has been outspoken about these challenges, and its behavior on the international stage pertaining to climate issues can be understood as an important component of the

Case 1: (cont.)

organization's broader efforts to achieve international legitimation. Its actions surrounding COP26 – the UN's annual climate change conference held in Glasgow, UK in 2021 – illustrates this impetus. Despite its exclusion from international climate treaties, the Sahrawi Republic voluntarily produced its own document mirroring the Nationally Determined Contribution (NDC), a climate action plan expected of all state parties to the Paris Agreement, and released its NDC report to coincide with the COP26 negotiations. In the fifty-seven-page report, the SADR commits itself to "the multilateral processes" established under the Paris Agreement and lays out a series of climate change mitigation and adaptation measures to be adopted conditional on external climate financing support (SADR Office of the Prime Minister, 2021). Polisario representatives publicly launched the report at a side event of COP26 known as the "people's summit," where various unrecognized states and civic organizations – i.e. those unaccredited for UN proceedings like COP – engage in a sort of parallel conference. There, and through international media, Polisario officials decried the Sahrawi Republic's exclusion from international climate governance mechanisms and the "climate injustice" such exclusion represents (Democracy Now, 2021; Western Sahara Resource Watch, 2021). The message Polisario sought to impart was clear: grant the SADR a seat at the UN as a recognized state and it will be a proactive player in international climate governance, representing a population facing severe climate-related vulnerabilities.

Rebel self-legitimation is often coupled with delegitimization efforts toward the state. Polisario thus also used the sidelines of COP26 to argue that Morocco was "greenwashing" its occupation of Western Sahara. According to Polisario officials, Morocco was positioning itself at the forefront of climate governance in North Africa and committing itself to increased renewable energy production on the one hand, but using the territory of Western Sahara for renewables including solar and wind power, without consent from the Sahrawi people, on the other (Democracy Now, 2021; Smith, 2018; Western Sahara Resource Watch, 2021). The case of Polisario thus demonstrates how armed groups seeking international legitimacy use environmental governance both to further advance their credentials as would-be states and as a platform to engage in yet another form of political contestation against the state.

CASE 2: REPUBLIC OF ARTSAKH (NAGORNO-KARABAKH)

With the dissolution of the Soviet Union in 1991, Armenia and Azerbaijan became independent states. Conflict between the two states began immediately over contested territory known as both the Republic of Nagorno-Karabakh and the Republic of Artsakh. The Republic of Artsakh was primarily composed of ethnic Armenians, but its territory was formally claimed by Azerbaijan when the latter gained independence. The Republic of Artsakh declared itself independent from Azerbaijan in 1991 as an initial step to join Armenia, but failed to gain international recognition. Over time, the Republic of Artsakh maintained strong links to Armenia and received military and nonmilitary support from their co-ethnics in Armenia and the diaspora. In 1997, a ceasefire agreement ended large-scale violence in the conflict. However, small-scale conflict continued with a significant increase in violent action after 2005. This culminated in 2020 with a major Azeri offensive through which Azerbaijan gained control over large portions of the territory previously controlled by the Republic of Artsakh, followed by another offensive in 2023 that resulted in the dissolution of the Artsakh government. UCDP codes the Republic of Artsakh as active in conflict (i.e. generating twenty-five battle deaths or more in a year) from 1991 to 1998, 2005 to 2008, and 2012 to 2022.

Similar to Somaliland, during its existence as a de facto state the Republic of Artsakh created significant local institutions, some of which were charged with the management of environmental issues, including the Department of Environment and Natural Resources and the Assembly Committee of Infrastructure. Both bodies had a mandate over the governance of natural resources and nature preservation. Artsakh also had a Ministry of State Service of Emergency Situations, responsible for adaptation, mitigation and hazard response (Gharibian, 2021). Its ministry of agriculture initiated a land reform program to transition away from a Soviet-style economy shortly after declaring independence (Sharrow, 2007). However, the most important aspect of environmental governance in this conflict revolved around two issues – water governance and the direct environmental impacts of the conflict. Both issues became pivotal points of tension in the conflict as each side used these issues to legitimize its position and delegitimize its opponent.

Water governance has become an urgent issue in the Caucasus as water has become increasingly scarce due to environmental and development pressures

CASE 2: (cont.)

and uneven distribution of water resources between different states and territories. The Republic of Artsakh was relatively water-rich, as it controlled four significant dams and thirty-six hydroelectric plants, including the major Sarsang dam and the associated reservoir. These water systems were managed by the Artsakh Republic, often in cooperation with the Armenian government (Ahmadi et al., 2022; Gharibian, 2021; Zanatta & Alvi, 2024).[45]

The Azerbaijani government frequently alleged that the Republic of Artsakh's mismanagement of the reservoir was an act of eco-terrorism as it led to floods and the withholding of potable water for the downstream territory under Azerbaijani rule. This claim of mismanagement was supported by reports from the Parliamentary Assembly of the Council of Europe in 2015 and 2016. Armenian and Artsakh authorities contested these reports, citing the committee's refusal to visit the area and accusing the Azerbaijan government of withholding water from them, as the canal of the Terter River that feeds the reservoir is under Azeri control (CEOBS, 2021). While cooperation on these issues would likely have led to a more sustainable and efficient water policy, Azerbaijan refused to cooperate with the Republic of Artsakh as doing so would, in their view, validate their claims of independence. At the same time, the Republic of Artsakh used cooperation attempts to raise awareness of the perceived wrongs committed by Azerbaijan and to gain international recognition of its governance project. These cooperation attempts were successful in gaining international goodwill. For example, the 2013 proposal of the Republic of Artsakh to co-govern the Sarsang Reservoir and the Tartar River that feeds it gained the support of the Minsk Group – an International Organization created to deal with the Armenia and Azerbaijan conflict. Ultimately, this was rejected by Azerbaijan (Leylekian, 2015; Zanatta and Alvi, 2024), and Azerbaijan succeeded in preventing the participation of Artsakh in regional water governance projects, forcing Artsakh to rely on Armenia for representation (Waisova, 2017).

Figure 9 shows the locations of relevant dams and fires. The climate component of the conflict became increasingly relevant internationally in the conflict's later stages. Specifically, both parties made allegations of ecocide and eco-terrorism against the other due to the water governance

---

[45] After the 2020 conflict, only six hydroelectric plants remained under Armenian-Russian-Artsakh control. However, this included the major Sarsang reservoir.

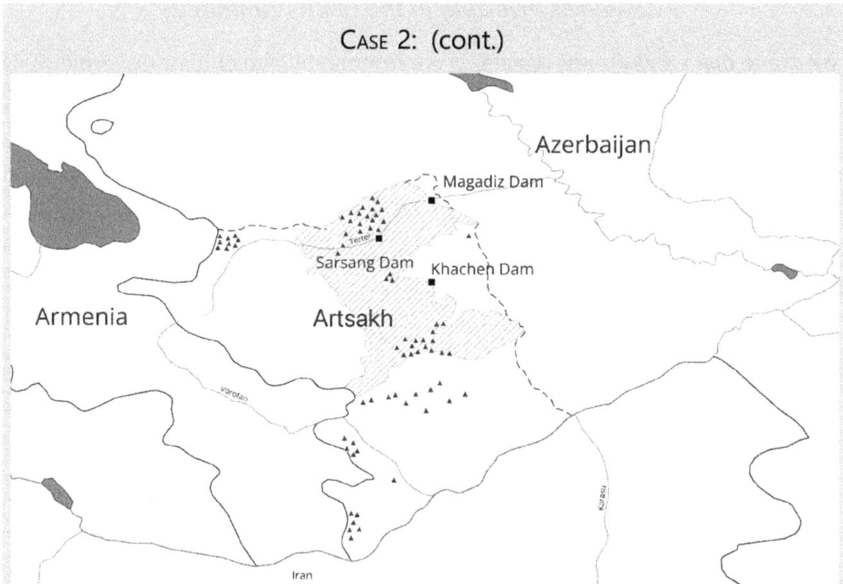

Figure 9 Republic of Artsakh with context dams and 2020 fire hotspots indicated

issues and the use of incendiary ammunition and tactical fires by one or both sides, which caused large-scale deforestation in the region (CEOBS, 2021; Zanatta and Alvi, 2024). This culminated with large disinformation and propaganda campaigns by both sides aimed at blaming environmental destruction caused by the conflict on their opponent. These campaigns mostly took place on the internet and involved actors such as local and international NGOs, the government of both Azerbaijan and the Republic of Artsakh, the public, and influential people such as K-pop artists, the metal band System of a Down,[46] and Kim Kardashian (CEOBS, 2021). While no independent research has attributed blame for the observed environmental damage from the conflict to either party, both parties used the issue of environmental damage to legitimate their position on Artsakh independence.

---

[46] The group also made a song/video Protect the Land which includes themes of alleged environmental crimes by Azerbaijan as part of a fundraising initiative. Artsakh Public TV is listed under the Special Thanks category for the video.

4.2.2 Rebels Investing in the Local Community

We argue that rebel groups engage in environmental and climate governance as a form of social welfare; in the age of climate change, they seek to ensure that the territory over which they stake a political claim is productive and sustainable. We examine this dynamic in the cases of the Moro Islamic Liberation Front in the Philippines and Somaliland's unrecognized government, highlighting the ways rebels implement environmental projects to cater to the livelihood needs of local communities and territories.

CASE 3: MILF IN THE PHILIPPINES

The Moro Islamic Liberation Front (MILF) traces its origins to the Moro National Liberation Front (MNLF). The MNLF fought for the independence of the Bangsamoro people of the southern Philippines and rebelled against a repressive government and an influx of Christian Filipinos to their traditional homelands. Peace talks between the MNLF and the government led to splinters in the movement. A faction led by Salamat Hashim fully broke with the MNLF to form the MILF in 1984. UCDP codes the MILF as active in conflict (i.e. generating twenty-five battle deaths or more in a year) in 1990, from 1994 to 2011, and in 2015.

The MILF distinguished itself from the MNLF in its stronger emphasis on an Islamist ideology and its refusal to accept any outcome short of independence. Both of these factors influenced the group's approach to governance broadly, including over the natural environment. The MILF believes that sustainability and self-sufficiency are interlinked, and that self-sufficiency is necessary to achieve self-determination (Taya, 2007). The territory claimed by the Bangsamoro people on the southern part of the archipelago is at high risk due to climate change, as rising sea temperatures threaten food and security, as well as coastal ecosystems (Ewing, 2009).

The MILF engaged in a lengthy peace process with the Filipino government starting in 1997, even as the conflict continued at low intensity until 2015. One landmark in this process was the Joint Statement of Kuala Lumpur in 2003. This process was pivotal for the MILF's governance project as it fully implemented the Bangsamoro Development Agency (BDA), which would become the central governance achievement of the MILF until later implementations of the Bangsamoro Organic Law (BOL) in 2018 (Bangsamoro Development Agency, 2015; Taya, 2007).

Throughout its existence the MILF set up a robust governance system in its territory. MILF governance can be divided into two phases: the pre-

CASE 3: (cont.)

BDA phase before 2001 and the post-BDA phase thereafter. In the first phase, the MILF provided governance directly through its political structure with Province and Barangay (local) level committees that presided over issues concerning the environment. The MILF directly engaged in farming activities, such as growing corn in their camps for local consumption and charity. The MILF also provided subsidies, financing, agricultural materials such as pesticides, fertilizers, irrigation, tractors, and no-interest loans based on Islamic Law, and often wrote off debts for those in their areas of control when there were crop failures. In exchange, farmers receiving support were expected to sell their products, often at fair market prices, directly to MILF dominated cooperatives (Bangsamoro Development Agency, 2015; Taya, 2007).

While the MILF often pushed for the modernization of agriculture (Taya, 2007), they also protected traditional practices as observed in the Ligawasan marshes. Local fighters cooperated with traditional leaders and governments to enforce indigenous fishing and rice farming laws in the Ligawasan marshes. These traditional practices include observing farming beliefs/rituals (often associated with greater sustainability (Sinolinding et al., 2013)), banning electric and chemical fishing and fly-catching, and open access regarding the rights to control accessibility and utilization of the marshland. Between 1990 and 2000 aquatic wildlife such as fish, crustaceans, and mollusks remained abundant in the marshes (Sinolinding et al., 2013). During this period, the MILF also worked directly with the Filipino government – despite the ongoing conflict – on infrastructure projects such as the multimillion-dollar Maridagao-Malitubog Irrigation project in northern Cotabato, which began its planning phases in 1995. The MILF gave the Filipino government access to areas under its control for the construction of water infrastructure, including inside its largest camp, Abu Bakre As-Siddique (Department of Environment and Natural Resources, 2003; Taya, 2009). Figure 10 shows the territory of the Bangsamoro Autonomous Region in Muslim Mindanao (BARMM) and governed marshlands.

Initially, the MILF engaged in minimal cooperation with INGOs, foreign states, and IOs on environmental issues, as the organization believed reliance on outside actors would make it vulnerable to pressure from donors. But the creation of the BDA drastically changed MILF governance and its approach to outside support. The BDA is the development wing of the MILF movement that (until 2018) acted as a governance institution (Bangsamoro Development Agency, 2015; *BDA Profile*,

Figure 10 The BARMM territory with relevant marshland highlighted

n.d.; Walsh et al., 2018). The BDA was a unique development agency, as it was controlled by a non-state armed group but was recognized by the government and the international community as a fully operational development agency. The establishment of the BDA signified an important achievement for the MILF; the group could receive foreign funds and manage them without government interference while enjoying a stamp of approval from the international community and the government, even during periods of conflict (Taya, 2009).

With growing international support, the BDA expanded the MILFs policies of agricultural modernization and self-sufficiency, sustainability, and aid. It created an Environmental Protection office, as well as other offices that implemented development projects, in attempts to uphold sustainable and environmentally friendly practices (Bangsamoro Planning and Development Authority, 2020). These offices were responsible for numerous development projects that dealt with agriculture and water management, including the Socio-economic Reconstruction and Development in Conflict-Affected Areas in Mindanao (SERD-CAAM) Project in 2006 and the Mindanao Trust Fund-Reconstruction and

Case 3: (cont.)

Development (MTFRD) Phase 1 from 2006 to 2015. IOs, such as the World Bank and UNICEF, foreign NGOs such as the Mantana Foundation, foreign state backers such as Australia, Turkey, and Japan, and domestic civil society organizations cooperated with the MILF to implement these projects (*BDA Profile*, n.d.; World Bank, 2007). Language related to environmental protection, preservation, and sustainability was heavily featured in the planning and discussion of all BDA projects (World Bank, 2007).[47]

The BDA also served as the central coordinating unit for several disaster responses (UNHCR, 2012; Walch, 2014, 2018). For example, the International Organization for Migration (IOM) provided the BDA with technical assistance in humanitarian response and preparedness and undertook a "community-based interactive community mapping and installation of community maps identifying hazard areas, safe zones, and vulnerable populations as a mechanism to guide populations to safety in the event of future armed conflict or natural disaster" (IOM, 2014, p. 78). Another area of concern for the MILF and BDA is the resettlement of thousands of displaced Bangsamore due to conflict, climate issues, and land grabbing primarily conducted by Christian Filipinos who have migrated to the region for better economic and environmental conditions (Taya, 2009). Programs such as the previously mentioned MTFRD also include significant aid for IDPs (World Bank, 2007).

The MILF has tied its governance policy to its goal of independence. As such, it implements policies and programs aimed at sustainability to foster self-sufficiency and prepare the Bangsamoro people for independence. The group's founder Salamat Hashim believed that fostering self-sufficiency was critical if the MILF was to survive and succeed, as it would not become beholden to external actors. He further believed self-sufficiency and sustainability would allow the MILF to gather support to tackle issues faced by the Bangsamoro people, such as poverty, underdevelopment, and the migration of Christian Filipinos.

---

[47] For example, the Bangsamoro Development Plan (BDP), written by the BDA in 2015, outlines the environmental strategy for the Bangsamoro autonomous region and stresses the need to develop the Strategic Environmental Management Plan (SEMP) and the Bangsamoro Waters and Zone of Joint Cooperation between the central government and the Bangsamoro government (Bangsamoro Development Agency, 2015).

Case 4: Republic of Somaliland in Somalia

Somaliland proclaimed its independence from Somalia in 1991 following a devastating civil war against the regime of Siad Barre. Since then, the Republic of Somaliland has functioned as a universally unrecognized de facto state with a highly institutionalized governing structure. It maintains a civil administration, a police force, democratically elected representatives, and a taxation system, and provides basic public services. UCDP codes the Republic of Somaliland as engaged in active conflict (i.e. generating twenty-five battle deaths or more in a year) in 2018. Nevertheless, Somaliland has continually pressed for independent statehood through political and diplomatic means throughout its existence as a de facto state. Figure 11 shows the current territory claimed by Somaliland.

From the outset, self-sufficiency has been a mainstay of Somaliland's claim to independence. The Somali National Movement (SNM), the rebel

Figure 11 Somaliland contested territory in Somalia

Case 4: (cont.)

group that fought against Barre's forces in Somaliland's independence war, drew on the support and productive capacity of Somaliland's local populace to fund its insurgent campaign; this "reinforced an ethos of economic and political self-reliance from an early age" in Somaliland (Phillips, 2020, p. 56). Indeed, Somaliland's sovereignty discourse draws prominently on its track record of democratic governance and self-sufficiency (Phillips, 2020, pp. 62–70).

With its predominantly agro-pastoral economy, Somaliland faces numerous adverse impacts of climate change including recurring droughts, floods, biodiversity loss, land degradation, and population displacement (Belay & Sugulle, 2011). These effects have secondary impacts on communities' livelihoods, health, food security, water access, and economic productivity (Omer, 2024). Amidst these threats, the Somaliland government has engaged in a series of actions over the years, largely with the support of international organizations and NGOs, towards mitigation and adaptation. Much of what we can glean about Somaliland's responses is institutional in nature, consisting of establishing government ministries and associated websites, issuing strategic plans, introducing laws aimed at environmental protection, and hosting workshops with international development organizations. There is a performative aspect in these initiatives that may be intended for international audiences, thus making this case consistent with our argument on rebel interests in gaining international legitimacy (on this, see also the cases of Polisario and Nagorno-Karabakh above). As Phillips writes, "Within Somaliland, there is studious attention paid to the international coverage that the country receives . . . . There is also a high level of knowledge among [political elites] about what is being said about Somaliland in the international media and academic spaces" (Phillips, 2020, p. 156). With the Somaliland government itself being the source of much of the information provided in this discussion,[48] one should be mindful of the self-legitimation interests behind these efforts. Nevertheless, it is also clear that the various initiatives collectively tell of a government deeply concerned about climate impacts on local

---

[48] International organizations often report on Somalia as a whole rather than on Somaliland specifically, even when projects are implemented in Somaliland proper. This makes data collection on Somaliland particularly challenging and reflects the fact that Somalia is represented in international membership bodies such as the World Bank and UN agencies while Somaliland is not.

CASE 4: (cont.)

communities and keen to partner with domestic and international actors to address them.

Somaliland boasts several government ministries whose work focuses on environmental governance. The Ministry of Environment and Climate Change (MoECC) leads the charge and oversees projects ranging from charcoal reduction, water management, and wildlife conservation to marine and coastal management. The MoECC partners with numerous external organizations including the World Bank, UNDP, World Vision, the Danish Refugee Council, and Save the Children.[49] More sector-specific government entities include the Ministry of Water Resources and Management, the National Disaster Preparedness and Food Reserve Authority (NADFOR), the Ministry of Public Works, Land, and Housing, and the Ministry of Agricultural Development.

While institutional developments are more easily observed than programmatic details and impacts, reports by aid organizations document the more specific environmental governance projects implemented in cooperation with the Somaliland government and local communities. These include water infrastructure development, disaster management planning, rainwater harvesting, soil conservation, food security, and combating wildlife trafficking (Cheetah Conservation Fund, 2023; Sifuma, 2016; UNDP, 2016; UNDP Somalia, 2016).

Somaliland has also introduced a series of environmental laws. Most notable is the Environmental Management Law no. 79, introduced in 2018. Its 81 articles lay out principles of environmental management, including the preservation of cultural and natural heritage, and stipulate regulations for environmental governance and penalties for damages caused to the environment (Republic of Somaliland, 2018). The Somaliland government also convenes an annual "Somaliland Climate Change Conference" bringing together both local and international stakeholders, with sessions devoted to specific themes such as involving youths and the media in raising climate awareness and promoting sustainable rangelands.[50]

The Somaliland government is upfront about continued deficiencies in areas such as institutional capacity and coordination between various government programs (Republic of Somaliland, 2017), as also noted in

---

[49] See https://moecc.govsomaliland.org/.
[50] See MoECC Twitter posts at https://twitter.com/MOECC_JSL.

> CASE 4: (cont.)
>
> academic research (Omer, 2024, pp. 7–8). Nevertheless, though unrecognized and excluded from international climate conferences, Somaliland has taken a proactive stance on its climate crisis, most notably through the creation of bureaucratic and legal structures around environmental governance. The government is engaged in a series of actions aimed at encouraging environmental sustainability in the face of multiple climate-related threats and ensuring residents' livelihood needs are addressed. Given its de facto self-rule and independence objectives, the Somaliland government is duly aware that it must devote efforts to local climate action and environmental protection to safeguard its populace and its territory.

4.2.3 Rebels Seeking Tactical and Strategic Advantage

Environmental governance can also serve rebel groups' military objectives, sometimes at the expense of environmental protection itself. The cases of Myanmar's KIO and the Islamic State in Syria demonstrate the sometimes fine line between environmental protection and the weaponization of the environment for tactical and financial gain.

> CASE 5: THE KACHIN INDEPENDENCE ORGANIZATION (KIO) IN MYANMAR
>
> The Kachin Independence Organization is one of many organizations that emerged to seek greater self-rule in Burma after Burmese independence. While the group began fighting for independence in the early 1960s, they changed their aim to autonomy within the Burmese state and signed a ceasefire agreement in 1994. The ceasefire ended in 2011, and violent conflict has resumed. UCDP codes the KIO as active in conflict (i.e. generating twenty-five battle deaths or more in a year) in 1961–1992 and 2011–2022.
>
> The KIO, and their affiliated Kachin Independence Army (KIA), have a complex and multifaceted relationship with the environment. As a region that obtained a measure of autonomy from the central government (through force and accommodation), the KIO has been able to establish a degree of territorial control and administration in the region. This includes governance activities in a range of areas, including establishment of land, forest, and natural resources policy (Woods, 2019). This long running civil war highlights the potential for tactical and strategic advantages of the natural environment, but also how quickly incentives to protect can shift to exploitation as conflict dynamics shift.

CASE 5: (cont.)

The Kachin area is in the northern part of Burma on the border with China and India, and is heavily forested (in 2010, over 85 percent of the Kachin territory was forested).[51] The KIO rebels initially benefited significantly from the dense forest cover (Sadan, 2015). In the early years of the insurgency, the KIO headquarters was based in a mountain fortress in Pa Jau. It moved its headquarters to a small border town (Laiza) on the Chinese border at the signing of the 1994 ceasefire agreement (Brenner, 2019, p. 76). Figure 12 shows the Kachin State and highlights an area where the KIO has banned mining projects.

The 1994 agreement did not address underlying political issues per se (Oo & Min, 2007), but was part of an overarching strategy by the military government in Rangoon to manage threats to the state. This transition to a

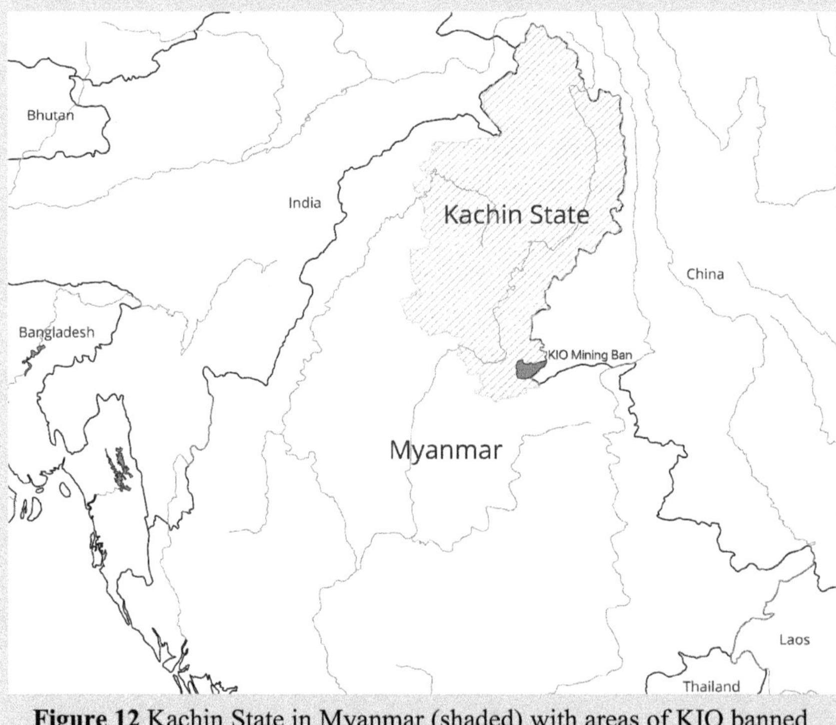

Figure 12 Kachin State in Myanmar (shaded) with areas of KIO banned mine (solid black)

---

[51] www.globalforestwatch.org/dashboards/country/MMR/4/?category=undefined&map=eyJjYW5Cb3VuZCI6dHJ1ZX0%3D.

CASE 5: (cont.)

period of ceasefire moved the KIO into a period of coexistence with the state (Callahan, 2007). The KIO was able to develop both means of social control, including many that address environmental issues, and extractive capacity (especially logging). The organization engaged in taxation, as well as issuing land tenure certificates (Hong, 2017). Their relationship with forest cover shifted from protective to highly extractive at this point.

Post-ceasefire, we see evidence of deforestation in the KIO areas as the rebel group extracted timber for profit. In an interview with a local informant after the ceasefire, researcher David Brenner is told "We know that it's not good for the environment, and environment agencies prohibit timber cutting like that, but we need to manage our income. So we need to do it. We need that business" (Brenner, 2019, p. 81). While the group has previously been able to extract funds from mining jade, this option became untenable. Kevin Woods quotes a KIO official as saying that "after the ceasefire agreement they took the trees instead" (Woods, 2011, p. 15).

This case demonstrates the intersection of complex goals and motivations of rebel actors with respect to the natural environment. Environmental protection can offer tactical advantages in conflict, as argued, but so can environmental destruction through activities such as logging and mining. The KIO benefited from and initially protected forests at the outset of conflict. It subsequently exploited them when need and opportunity arose. In 2002, the KIO "declared it illegal to log without their explicit permission" and government approval (Woods, 2011, p. 17). In 2003, the group canceled a mining deal intended to fund both local development and weapons acquisition in direct response to environmental concerns expressed by the local population (Fishbein et al., 2023). The KIO has also encouraged and supported the establishment of community forests in areas they control, which are a key part of sustainable forest management (Nalbo, 2019, p. 107). Forest management in the past decade in the Kachin state has included a multitude of actors, and the KIO has been playing an active role in this space (Nalbo, 2019).

CASE 6: ISLAMIC STATE IN SYRIA

The Islamic State (IS) emerged in Iraq around 2004 as an al-Qaeda ally. By 2013, the group had broken its relationship with al-Qaeda and established itself as the Islamic State, having taken control of substantial territory in

CASE 6: (cont.)

Syria and Iraq and operating as a quasi-state within its zones of control (Gerges, 2016). UCDP codes the Islamic State as active in conflict (i.e. generating twenty-five battle deaths or more in a year) from 2013 to 2022.

IS's complex governance structure in its territories functioned as a government with federal, regional, and local organizations. This bureaucracy oversaw the governance of the economy, agriculture, and social and religious life, and provided public goods such as emergency aid, water management, and migration assistance (Bamber-Zryd, 2022; Caris & Reynolds, 2014; Ingram et al., 2020). IS also cooperated with local organizations in their provision of aid and agricultural assistance. These were primarily local councils (civilian-led institutions that emerged during the civil war and typically coordinated with, but operated separately from, armed groups) as well as tribal networks and Islamic-aligned civil society groups (Carnegie et al., 2022; Khalaf, 2015).

One of the main issue areas for the IS was managing Syria's water through the many dams it controlled. Dams are pivotal climate governance infrastructures in any region, as they permit whoever controls them to provide or withhold water and electricity to significant portions of the local population. By controlling these visible environmental infrastructures and providing essential services, actors can demonstrate capacity and generate legitimacy for their rule (von Lossow, 2016). This is especially true in Mesopotamia, an area that faces chronic water shortages due to mismanagement and changing climate conditions and where most people rely on water distributed by the region's many dams to feed the agricultural economy. For example, IS used dams in Tabqa in Syria to impose its control over agricultural production – the region's main economic driver – displaying bureaucratic capacity that matched, if not exceeded, that of the Syrian regime. This allowed them both to capture the region's resources and to increase their local legitimacy (Daoudy, 2020). Because of these benefits to controlling the natural environment, IS made the capture of these sites a strategic objective and incurred heavy losses to gain control of almost all of the dams in Syria and Iraq by 2015 (von Lossow, 2016). Figure 13 shows the dams that have been sites of contested control and governance in the conflict.

CASE 6: (cont.)

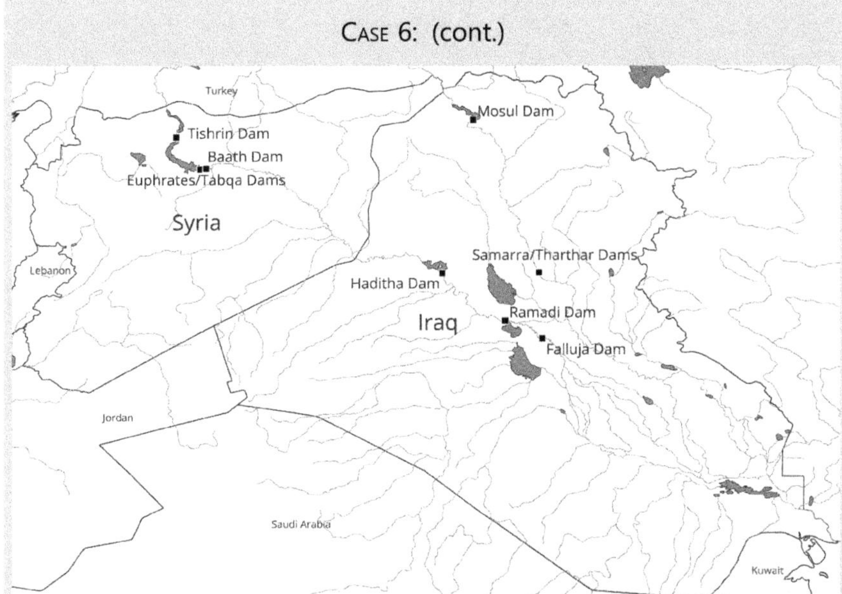

Figure 13 Dams at times controlled by ISIS in Syria and Iraq

Besides the governance benefits of dams, these environmental infrastructure pieces provide significant tactical advantages that encourage groups such as IS to control and operate them. IS used dams as a weapon by depriving some regions of water, thus creating man-made droughts to damage the official government's authority and weaken local resolve, while flooding other areas to cause large-scale damage and prevent military operations (Daoudy, 2023; von Lossow, 2016). For example, in April 2014, IS released the floodgates to the Fallujah Dam in Iraq (after having closed the floodgates to stock water), causing massive floods that destroyed over 10,000 homes, 200 square kilometers of farmland, killed most of the local livestock and left 60,000 people displaced (von Lossow, 2016). Even when actors refrain from unleashing the destructive capacity of dams, the threat of doing so creates tactical difficulties for others engaged in the conflict and can be a powerful bargaining tool.

Dams also make natural strongholds, not only due to their sturdy concrete and earth architecture, but due to their centrality in governance. Dams are sources of electricity and water that have the potential to cause major environmental disasters if damaged. As such, all parties to a conflict face a simultaneous incentive to gain control over dams while refraining from using weapons and tactics that can cause critical damage to them.

Case 6: (cont.)

These incentives are visible in the conflict over Syria's largest dam in Tabqa. Because of the dam's importance and fears of a major catastrophe if damaged, the dam was one of the sites on the United States's "no-strike" list. American and allied forces were restricted to limited strikes using up to 20-pound warheads on the earthen sections of the dam and prohibited from using any bombs or missiles, no matter the size, on the concrete structures that controlled water flow (Philipps et al., 2022). These self-imposed limits on American force use in the conflict greatly benefited IS, as it could continue to operate the dam as part of its governance project and as a relatively safe base of operations.

Despite the Tabqa Dam being on the no-strike list, American forces eventually used large explosives in an attack on the dam in January 2017. These strikes damaged critical equipment and infrastructure and impinged on the dam's operations. The dam reservoir quickly reached capacity and faced an imminent threat of overflowing, leading to a truce and a rare moment of cooperation between IS, the Syrian government, the SDF, the US, and Turkey. As the probability of a breach in the dam increased, Turkey cut off water flow to the region using its dams upstream to buy time as a team of sixteen engineers from the Syrian government, IS and the SDF under a temporary ceasefire worked to get the dam to minimal operational levels to prevent a large scale catastrophe (Daoudy, 2020; Philipps et al., 2022). This ceasefire again benefited IS forces, providing the group time to prepare for the eventual ground attack by the US-allied SDF on the dam. The case of the IS in Syria shows how tactical and strategic goals intersect with rebel protection of the natural environment.

4.2.4 Interpretation of Findings from Cases

The purpose of the vignettes is to bring to life the variance and multidimensionality of rebel environmental governance. REG+ varies in the degree of formality and institutionalization spanning more or less ad hoc forest protection by the KIO to a full-blown state-sanctioned rebel development agency run by the MILF. It varies in form, ranging from the management of dams by the IS in Syria to the Polisario staging a presence at an international climate summit. It also features different levels of cooperation with state authorities and international bodies. While motives are multifaceted, the vignettes provide suggestive evidence for the arguments presented here on rebel efforts to ensure sustainability for local communities, seek domestic and international legitimacy, and realize tactical gains. Yet, they also show how rebel interests in protecting the environment are intertwined with their pursuit of military objectives and even financial profits to fund their insurgencies, underlining their complex drivers.

5 Conclusion and Implications for Further Research and Policy

5.1 Summary of Findings

This Element has brought together quantitative and qualitative data on rebel engagement with the environment in the age of climate change. It marshaled novel data to provide the first systematic quantitative and qualitative evidence that rebel groups are active agents in governing over environmental issues. From building wells and regulating land usage to creating and maintaining nature preserves, rebels do so in numerous and often inventive ways. While very small in scale, some of these activities are consistent with climate mitigation activities, namely preserving forest and land. As climate impacts mount, this governance also extends to include climate adaptation to maintain agricultural activities and other services. Rebels are also providing relief services following climate-induced hazards and taking steps to increase resilience. Thus, rebels are not only responding to the changing conditions on the ground driven by climate change; they are proactively addressing current and future hazards that are associated with climate change in a variety of contexts.

In a small number of cases, rebel groups have been willing partners of the international community in addressing climate change, holding meetings on the sidelines of international summits and entering into dialogue with international organizations and NGOs about climate mitigation and adaptation efforts. Both Polisario (representing the Sahrawis in Western Sahara) and an ULMWP affiliate of the Free Papua Movement (OPM) in Indonesia have attempted to engage directly with COP meetings (SADR Office of the Prime Minister, 2021; UDCP, n.d.; ULMWP, 2021).

Furthermore, rebels are increasingly *talking* about climate change and the environment. Rebel groups that started in the mid-2000s and after talk publicly about these issues at a much higher rate than those in the 1990s and early 2000s. Whether or not they are taking environmental action, they are embracing the topic of climate change as part of their discourse, demonstrating their understanding that local issues they face are linked to global concerns and that climate change is relevant for their strategic framing. They may or may not be welcomed at international climate summits, but they are certainly inserting themselves into global climate conversations.

Under what conditions do rebels engage in environmental governance? We leveraged our novel REG+ data to highlight patterns of governance globally in civil wars fought over territorial claims. A preliminary examination of this data, in combination with data on a number of important features of civil wars, reveals a few trends that align with our empirical expectations, but also open up new avenues for future scholarship in this area.

Our analysis lends preliminary credence to our theoretical assertions: Rebels that are more invested in the local community, and those that are seeking domestic and international legitimacy, appear more likely to use REG+, as are rebels who see tactical or strategic advantages to environmental governance. We provide six qualitative vignettes which explore the provision of governance in more detail, highlighting the motivations that drive rebel behavior. Literature on rebel governance broadly tells us that rebels govern when they are invested in the local community and seek legitimacy. We see this in the cases of environmental governance as well. Rebel actors are, at times, explicit about the need to protect the environment for the population they hope to fully govern. Rebels also engage in a variety of actions, such as informally opting in to international treaties, to garner legitimacy. Similarly, we observe rebels calling attention to their efforts to manage and preserve the environment on the international stage. A key finding from the qualitative vignettes is the complexity of rebels' relationship with governing issues related to the environment. Similar to many states, rebel actors sometimes work to manage and protect the natural environment while also engaging in practices that degrade the environment (such as mining and logging).

5.2 Future Research

We situate our work within broader calls to examine climate change through the lens of appropriate social science disciplines (Schipper et al., 2021). Specifically, this research addresses the challenges that multiple scales of efforts pose for social science research (IPCC, 2023a), particularly with respect to exploring local decision-making by a variety of actors at different levels of impact (Dzebo & Stripple, 2015). Additionally, our work responds to calls to develop new data for contexts which have been less well explored, namely areas affected by armed conflict, fragility, and high levels of vulnerability (Sitati et al., 2021). Our work points to important directions for further inquiry for political scientists and for interdisciplinary research agendas on climate change.

First, much room remains for further research at the intersection of rebel governance, various aspects of environment and climate governance, and broader questions related to governance and contestation. One key question is how rebel environmental governance is similar to or different from other governance efforts. Existing scholarship has focused on issues like education, health care, and security provision. Rebel governance is at least in part about providing for a local population. Yet, the payoffs of some REG+ activities have a longer time horizon and less clear payoff for rebels in terms of generating local compliance or even local welfare in the short term. Our findings here based on analysis with the Quasi State Institutions data suggest that the provision of environmental governance aligns closely with other efforts to create governance structures and legitimize these efforts internationally. It appears less clearly connected to rebel efforts to provide basic services. Future work could usefully explore rebel decision making about REG+ activities directly, and contextualize this behavior in light of other types of governance provided by these groups.

Additionally, future research should examine how state environmental governance interacts with rebel efforts. Do the same factors that help explain variation in state environmental governance also help explain variation among rebels? State neglect or failures in environmental governance can motivate rebels to step up their efforts to paint a contrast, as in Somaliland, but state achievements, or even rhetoric, could also motivate rebels to match their efforts. The study of the dynamic interplay between state and rebel environmental governance will be facilitated by further advances in understanding why states themselves vary in their efforts to preserve the physical integrity of their territories.

Another key question for rebel governance scholars relates to the effect of these behaviors on rebel success. We theorize rebel environmental governance as a strategy to advance rebel aims. How does the provision of REG+ influence the chance of negotiations, settlement, or external support in a dispute? Rebels strategically frame their behavior for internal and external audiences, but we have little understanding of the returns to rebels for engaging in REG+, or what factors might condition the efficacy of attempts to market themselves as "eco-rebels."

Moreover, the work raises questions about the dynamics of rebel group cooperation with IOs and nongovernmental organizations. Studies have shown that cooperation is possible between rebels and the states they violently challenge in some instances (Beardsley & McQuinn, 2009; Schievels & Colley, 2021). Less is known about the conditions that favor cooperation with other types of actors. Greater attention should be given to the occurrence and nature of cooperation between rebels and international organizations more broadly.

This work also provides an opportunity to dive into variation in governance in an issue domain that will contribute to the accumulation of knowledge, particularly in fragile and contested spaces. The REG+ data can be used to explore, for example, core questions in the study of governance, such as the differences between rule-making, institution-building, and service delivery. Additionally, the REG+ data will add new leverage to questions around the dynamics of cheap talk and publicity, examining the links between rebel use of environmental rhetoric and rebel actions (or lack of actions) on the ground to manage, protect, or degrade the environment.

Second, our work raises a series of questions related to authority, responsibility, and the politics of blame. Research suggests many citizens broadly understand the causes of climate change, with some important variation (Druckman & McGrath, 2019; Simpson et al., 2021). Yet, we have little understanding of how people attribute responsibility for climate impacts and the resulting harms, and who they hold responsible for preventing and moderating these damages. There are several challenges in this line of inquiry. Even with a relevant understanding of climate change, not all events or observed changes can be linked with any certainty to climate change; the impacts are always a combination of the hazard, exposure, and vulnerability, which is in turn modified by adaptation and development (Muller, 2018). Thus, blame can be as much an issue of perception as reality. Experts agree that developed countries are responsible for much of the overall harms due to historical emissions and that this should be reflected in climate finance arrangements to developing countries (Otto, 2017; E. Williams, 2020). People may still hold their own government responsible for harm or inaction (Nalau et al., 2015) or be swayed towards authorities that show more attention to aspects of their wellbeing that are being affected by climate change. The possibilities for this issue to be manipulated as both a way to assert authority or to pass blame presents pressing research questions.

Relatedly, the diffuse and varied effects of climate-induced hazards, from floods to droughts to increasing heat, provide many possible flashpoints for citizen demands and levels at which action might be taken. Future work should address how efforts by violent and nonviolent non-state actors can shift authority away from the state. Particularly pressing is the question of how the existence of hybrid or overlapping governance affects citizen buy-in to local governors.

Finally, this research raises questions for political science within the broader climate science agenda, especially as it relates to climate adaptation, governance, and international cooperation. While climate change is still largely understood through its physical characteristics, perceptions of these changes, the capacity to manage the impacts, and the potential for adaptation to avert the worst effects are coming to the foreground, including the need for governance approaches at multiple scales (Brondizio et al., 2016). By identifying key aspects of local governance conditions in conflict-affected settings, our work opens up new opportunities for engagement at multiple scales towards broader theories of governance to improve the overall understanding of human systems and the management of resources. This also points to new opportunities to expand work at the intersection of climate change, conflict, and security. Most research to date focuses on how environmental degradation affects conflict propensity. Our study suggests new research directions that consider the motivations of the rebel groups as mediating these conditions and thus a new space for research at the nexus of climate, security, and peacebuilding.

5.3 Policy Implications

The recognition that rebel groups are governing along multiple dimensions that relate to climate change raises a number of critical questions for current policy environments, including climate adaptation and finance, security, humanitarian organizations and their efforts in the context of climate-induced hazards, and peacebuilding. Here, we discuss some critical aspects that can be informed by our findings. There is a pressing need for more opportunities for climate adaptation in conflict affected spaces. Current approaches center on rebel groups primarily as producers of violence, overlooking the roles they might have in decreasing vulnerability, as well as how their conservation efforts can have positive spillover effects for climate change. Rebel groups present both challenges and opportunities for climate adaptation in these places as they mobilize around climate issues, including the possibility of engagement. At the same time, rebel groups are violent actors and all engagements with them should be carefully evaluated in light of this and the fact that there may be complex political and security implications at play. Yet, there may be opportunities for policy actions that can both meaningfully improve the state of climate adaptation in these vulnerable contexts as well as present novel opportunities for peacebuilding and conflict resolution.

5.3.1 The Imperative of Climate Adaptation and Finance in Conflict Affected Spaces

The IPCC AR6 Report emphasized the critical role of vulnerability as one of the root causes of climate risk, such that the more vulnerable are disproportionately affected. We join others who call for attention to climate action, especially as it relates to climate adaptation, to focus on alleviating the vulnerabilities that make people most susceptible to adverse effects from climate hazards. Those who live in conflict affected spaces have been rendered even more vulnerable due to the ongoing set of conditions that characterize most violent contexts. This is then amplified by climate hazards and may, in some cases, prolong and trap these areas in conflict (Buhaug & Von Uexkull, 2021). The lack of investments and programming to address climate hazards in conflict-affected areas serves to compound these risks. However, this deficiency also reflects the reality that interventions that are not sensitive to conflict contexts can also serve to exacerbate the conflict (Medina et al., 2025).

The needs of climate adaptation for populations in conflict affected areas and the implications for peace and security has been of increasing interest to the international climate policy community. Notably, the COP28 UAE Declaration on Climate, Relief, Recovery and Peace (DCRRP) focuses on improving the capacity of countries and communities in fragile and conflict affected contexts toward climate resiliency (COP28, 2023). This theme was continued at COP29 with the Baku Call on Climate Action for Peace, Relief and Recovery (COP29, 2024). A key pillar of this declaration is improving access to climate finance that has otherwise been undeveloped in these more vulnerable contexts (United Nations Development Programme, 2021; Venner et al., 2024). This increasing emphasis on serving these populations will present new issues for accountability and risk for funders, and creates a need to establish appropriate norms and metrics for implementation in these areas (Scartozzi, 2024) and for improved alignment of climate finance and other financial flows, including those related to peace and security (Läderach et al., 2021; Scartozzi et al., 2023; Wong, 2022).

These more positive concepts of improving delivery of finance and programming are also situated within sets of conflict dynamics. For example, issues related to co-opting this agenda also came to the forefront at COP29. The Baku Call on Climate Action for Peace, Relief and Recovery (COP29, 2024) was followed by concerns about how the nations that have also been aggressors in conflicts may seek to scope these activities towards their strategic interests with IGOs and NGOs in the climate, security, and humanitarian space. Several countries then signed on to a separate document, the "Common principles for effective climate finance and action for relief, recovery and peace" (Ecosystem for Peace, 2024).

This brings into focus the complex issues around delivery of climate finance and programming to support the most vulnerable where there needs to be as much awareness of the potential for exacerbating the conflict through rebel engagement as much as being co-opted by the nations who are the more conventional recipients of international or multilateral funding. Our findings inform these issues by presenting potential new actionable ways to more effectively manage these competing issues by better situating rebel groups within the dynamics of conflict and environmental governance. As most climate finance flows through national governments, the potential of co-option of these funds is more likely the cause of why conflict-affected states receive lower amounts of financing (Chambers & Kyed, 2024). This imperative for adaptation in these spaces along with the challenges posed by the current finance approaches has further highlighted the potential needs for separate funding approaches for conflict-affected regions and especially a willingness to emphasize grants to groups that would not be able to provide assurance of repayment. However, most of these proposals focus on finding and funding trusted partners, namely humanitarian and other international aid organizations, who are already on the ground to implement these efforts. The results in this Element highlight that at least in some cases, rebel groups could also through their engagement with climate-related activities have knowledge, capacity and relationships with the local populations on these issues that may entail consideration of whether and when to engage.

5.3.2 The Complex Climate and Security Implications of Rebel Engagement

If the international community is serious about the goals of climate adaptation and mitigation in conflict-affected territories, then concerned states and organizations *must* take up the question of how to engage with rebel groups operating in these areas. The default option to date – not engage at all, or engage only under very limited circumstances – will prolong the exclusion of the many non-state governors and their populations from climate deliberations, decisions, and actions, thus hindering progress toward the global objectives outlined in the COP28 declaration. It is, in essence, to continue to allow current politics to prevail over taking action to address climate change.

There are risks and challenges to engaging with rebel groups. Yet, where climate change intersects with conflict the stakes are high enough that stakeholders should invest the time needed to identify the opportunities and means through which engagement becomes possible. The benefits of engagement are straightforward. It means programming that addresses aspects of climate vulnerability, whether in the area of local security, social welfare, climate adaptation, disaster management, or food security, has a chance at implementation.

Where rebel groups are the governing authorities, rebel engagement is often a prerequisite for on-the-ground action. Our work has offered some explanations for why rebel groups engage in environmental governance. Given many rebel groups' skepticism toward external "meddling," external actors should capitalize on areas where their interests overlap with those of the rebels. Many rebel groups, as we have demonstrated, express a deep concern over the impacts of climate change on local sustainability and welfare.[52] Where this is the case, engagement should proceed on the basis of these shared interests. Where external actors see few overlaps with the concerns of rebel groups, engagement is likely to be more challenging, or rebel groups may reject outside interventions altogether.

International nongovernmental organizations (INGOs) operating in conflict-affected areas have demonstrated the success of rebel engagement (Okumu, 2003; Terpstra & Frerks, 2016). Groups such as the Center for Humanitarian Dialogue, the Carter Center, and Geneva Call have developed expertise in engaging rebel groups during armed conflict. These groups employ ad hoc contracts, other forms of agreements, and long-term cooperation to build trust and legitimacy with rebel groups (Hofmann, 2012). INGOs are able to act as a bridge in conflict settings to diffuse norms, build capacity, publicize abuses, and transfer knowledge to rebel groups distrustful of state actors (Hofmann, 2012; Stanton, 2020).[53] INGOs accomplish this through social pressure, direct incentives, and bargaining with rebel groups which often includes incentives such as the offer of technical support. While much of this engagement is centered around conflict mediation and violence reduction, similar techniques and similar organizations can be engaged towards the aim of enhancing climate adaptation and mitigation for vulnerable populations. Working directly with INGOs currently engaging with rebel groups governing over climate can advance these aims.

In addition, through engagement with rebel rulers, external actors can acknowledge and validate the concerns of local residents over climate-related issues. This is a first step toward enabling the most vulnerable populations to have a voice and broadening the climate discourse to include those who have so far been excluded from climate conversations at any level of governance.

The potential costs of engagement must also be acknowledged. Though they will vary depending on the context, one concern for external stakeholders is that rebel engagement becomes tantamount to rebel legitimization. This can have wider consequences. To approach, consult, and cooperate with rebel groups is to wade into the realm of politics because rebel groups are by definition opposing

---

[52] Others have identified rebel groups' varying commitment to the environment and potential avenues for engagement with international human rights law (De La Bourdonnaye, 2020).

[53] See also Matfess (2022).

the state with arms. Given this, external actors can face charges of bias or politicization of "aid" or "humanitarian" work whether by the state, its citizens, or by other observers (Abeytia et al., 2023; Wood & Sullivan, 2015). This is a continual and familiar challenge for any international organization navigating conflict-affected territories. A related consideration is whether and how engagement with rebels might affect the ongoing conflict. Legitimacy occupies a central place in rebel politics precisely because perceptions of legitimacy can strengthen rebels, helping them to garner popular support and recruit into their ranks, and can ultimately affect conflict dynamics. Interventions that bolster rebel legitimacy may thus have broader repercussions.

It is important to note that paradoxically, inaction, too, can result in rebel legitimization. One of the central arguments proposed in this Element is that rebel groups engage in environmental governance for its potential legitimacy payoffs. If external actors show inaction in the face of climate-related crises, rebel groups can rise to the occasion by positioning themselves as key providers of climate action while highlighting the hypocrisy of international climate discourses. These potential costs must be weighed against the potential benefits of engagement.

For many organizations involved in international development and humanitarian aid, these are perennial questions. How do they gain access to vulnerable populations in rebel-held territories while retaining their credibility with the state? How do they engage with rebels while "doing no harm?" It has now become a mantra that any external interventions into conflict-affected areas must exercise sensitivity to the local context and tailor their programming accordingly. Beyond echoing this call, this Element has highlighted a set of contextual considerations that are likely to be consequential for interventions at the intersection of climate and conflict: the extent to which rebel groups are involved with environmental sustainability and citizen welfare; the extent to which they demonstrate a concern for internal and external legitimacy; and the role of the natural environment in rebel tactics, from weaponization to conservation. As external stakeholders consider the broader political and social context, our work suggests these considerations can serve as a starting point for a more focused and grounded analysis.

None of this work is simple. It requires thought and planning involving political, legal, social, and ethical questions. The work is inherently multisectoral and multidisciplinary. Yet, ultimately, engagement with governing authorities at the subnational level – including with armed groups – will have to become routine if organizations are committed to climate adaptation and mitigation in areas affected by conflict.

5.3.3 Transforming the Landscape of Peacebuilding to Foster Synergies between Peace and Climate

This work also points to the need to transform systems to improve the structures in which climate action and peace are being linked and the associated policy landscape. As has been noted by others, many gains may be made by linking activities, such as climate and SDGs and Agenda 2030 synergies (Fuso Nerini et al., 2019). However, the interactions between climate action and development pathways (especially SDG16 and its goals on peace) are less well articulated in the literature and in practice than our understanding of the linkages between climate action and other environmental attributes. Specifically, the literature on the co-benefits of climate mitigation and adaptation for peace is still developing, although some examples include using environmental stewardship and governance of local common pool resources as a way towards local peacebuilding (C. G. Williams, 2024).

Here, climate mitigation policies come to the forefront as both essential for mitigating climate change, but also through their potential to provoke conflict (Gilmore & Buhaug, 2021). Mitigation strategies, particularly those focused on land use and carbon offsets,[54] may be of particular concern in settings that are conflict-affected. While these offset approaches are often presented as ways to generate funds in developing countries through afforestation and preservation of natural parks, they have also been associated with violence and conflict (Cavanagh & Benjaminsen, 2014). The state, in pursuit of financial gains, reduces local communities' access to land, especially where land tenure may be poorly enforced and reinforce marginalization. Recognizing that rebel groups – often through their territorial claims – are actively engaged in land conservation could introduce new avenues for linking climate policy and cooperation and peacebuilding. For example, mechanisms by which revenues and responsibilities for land-based mitigation are shared equitably could be discussed in conflict settlements. This could have the benefit of not only supporting climate policy but also fostering trust and producing revenues to support livelihoods in the post-conflict setting.

While rebel groups may at times be good governors, the violence that is inherent in these settings ultimately does not allow for most elements of Agenda 2030 nor climate action to be implemented, especially over longer durations. Regardless of the relative positioning of the rebel group and the state in these dialogues, resolving these conflicts remains the focus. However,

---

[54] It is beyond the scope of this text to also discuss the multiple issues with carbon offsets, including their current performance (West et al., 2020) as well as whether they are and how to make them consistent with stringent temperature goals (Cullenward et al., 2023).

peace settlements are often fragile in part due to structural issues that remain unaddressed in the settlements. For example, on gender, the UNSCR 1325 affirms that "peace and security efforts are more sustainable when women are equal partners in the prevention of violent conflict, the delivery of relief and recovery efforts and in the forging of lasting peace" (United States Institute of Peace, n.d.). As climate impacts are linked to entrenching vulnerabilities and increasing challenges for achieving the SDGs, placing climate actions as a central element in peace settlements could serve to reduce vulnerabilities and increase attention to pressing issues, thus fostering buy-in to agreements and further supporting the conditions in which international environmental agreements can be achieved.

References

Abeytia, A., Brito Ruiz, E., Ojo, J. S., & Alloosh, T. (2023). Do no harm: The role of humanitarian aid and neutrality in protracting civil wars. *Civil Wars*, *25*(2–3), 341–366. https://doi.org/10.1080/13698249.2023.2253101.

Abrahams, D. (2020). Conflict in abundance and peacebuilding in scarcity: Challenges and opportunities in addressing climate change and conflict. *World Development*, *132*, 104998. https://doi.org/10.1016/j.worlddev.2020.104998.

Aerts, J. C. J. H., Botzen, W. J., Clarke, K. C., et al. (2018). Integrating human behaviour dynamics into flood disaster risk assessment. *Nature Climate Change*, *8*(3), 193–199. https://doi.org/10.1038/s41558-018-0085-1.

Ahmadi, S. A., Hekmatara, H., Noorali, H., et al. 2022. The Hydropolitics of Upper Karabakh, with Emphasis on the Border Conflicts and Wars between Azerbaijan and Armenia. *GeoJournal* *88*(2), 1873–1888. doi:10.1007/s10708-022-10714-4.

Albert, K. E. (2022). What is rebel governance? Introducing a new dataset on rebel institutions, 1945–2012. *Journal of Peace Research*, *59*(4), 622–630. https://doi.org/10.1177/00223433211051848.

Allen, S., & Trinidad, J. (2024). *The Western Sahara question and international law: Recognition doctrine and self-determination.* Routledge, Taylor & Francis Group. https://doi.org/10.4324/9781032658827.

Amos, C. B., Audet, P., Hammond, W. C., et al. (2014). Uplift and seismicity driven by groundwater depletion in central California. *Nature*, *509*(7501), 483–486. https://doi.org/10.1038/nature13275.

Argus, D. F., Landerer, F. W., Wiese, D. N., et al. (2017). Sustained water loss in California's mountain ranges during severe drought from 2012 to 2015 inferred from GPS. *Journal of Geophysical Research: Solid Earth*, *122*(12), 10,559–10,585. https://doi.org/10.1002/2017JB014424.

Arjona, A. (2016). *Rebelocracy: Social order in the Colombian civil war* (1st ed.). Cambridge University Press. https://doi.org/10.1017/9781316421925.

Arjona, A., Kasfir, N., & Mampilly, Z. C. (eds.). (2015). *Rebel governance in civil war.* Cambridge University Press.

Asaka, J. O. (2021). Climate change – Terrorism nexus? A preliminary review/ analysis of the literature. *Perspectives on Terrorism*, *15*(1), 81–92.

Bäckstrand, K., Kuyper, J. W., Linnér, B.-O., & Lövbrand, E. (2017). Non-state actors in global climate governance: From Copenhagen to Paris and beyond. *Environmental Politics*, *26*(4), 561–579. https://doi.org/10.1080/09644016.2017.1327485.

Bamber-Zryd, M. (2022). Cyclical jihadist governance: The Islamic State governance cycle in Iraq and Syria. *Small Wars & Insurgencies, 33*(8), 1314–1344. https://doi.org/10.1080/09592318.2022.2116182.

Bangsamoro Development Agency. (2015). *Bangsamoro development plan.* Bangsamoro Development Agency. https://bda.org.ph/index.php/research-publication/bdp.

Bangsamoro Planning and Development Authority. (2020). *Bangsamoro development agency strategic plan 2020–2025* [Strategic Plan]. Bangsamoro Development Agency. https://iag.org.ph/images/pdf/1st-BDP-Final.pdf.

Barkhoda, D. (2016). The experiment of the Rojava system in grassroots participatory democracy: Its theoretical foundation, structure, and strategies. *Journal of Research in Humanities and Social Science, 4*(11), 80–88.

Barnett, J., & Adger, W. N. (2007). Climate change, human security and violent conflict. *Political Geography, 26*(6), 639–655. https://doi.org/10.1016/j.polgeo.2007.03.003.

BDA Profile. (n.d.). Retrieved April 26, 2024, from https://bda.org.ph/index.php/about-us/all-about-bda.

Beardsley, K., & McQuinn, B. (2009). Rebel groups as predatory organizations: The political effects of the 2004 Tsunami in Indonesia and Sri Lanka. *Journal of Conflict Resolution, 53*(4), 624–645. https://doi.org/10.1177/0022002709336460.

Belay, L., & Sugulle, A. J. (2011). *The impact of climate change and adoption of strategic coping mechanism by agro-pastoralists in Gabiley region, Somaliland.* Candlelight for Health, Education & Environment (CLHE). https://ke.boell.org/sites/default/files/the_impact_of_climate_change_in_gabiley.pdf.

Bergholt, D., & Lujala, P. (2012). Climate-related natural disasters, economic growth, and armed civil conflict. *Journal of Peace Research, 49*(1), 147–162. https://doi.org/10.1177/0022343311426167.

Berrang-Ford, L., Biesbroek, R., Ford, J. D., et al. (2019). Tracking global climate change adaptation among governments. *Nature Climate Change, 9*(6), 440–449.

Biehl, J. (2014). *Ecology or catastrophe the life of Murray Bookchin.* Oxford University Press.

Biesbroek, R., & Lesnikowski, A. (2018). The neglected dimension of polycentric climate governance? *Governing Climate Change Polycentricity in Action?*, 303–319.

Blumont. (n.d.). *Syria.* Blumont. Retrieved October 10, 2023, from https://blumont.org/where-we-work/syria/.

Bob, C. (2005). *The marketing of rebellion: Insurgents, media, and international activism.* Cambridge University Press.

Börzel, T. A., & Risse, T. (2021). *Effective governance under anarchy: Institutions, legitimacy, and social trust in areas of limited statehood* (1st ed.). Cambridge University Press. https://doi.org/10.1017/9781316872079.

Braithwaite, J. M., & Cunningham, K. G. (2020). When organizations rebel: Introducing the foundations of rebel group emergence (FORGE) dataset. *International Studies Quarterly, 64*(1), 183–193. https://doi.org/10.1093/isq/sqz085.

Brenner, D. (2019). *Rebel politics: A political sociology of armed struggle in Myanmar's borderlands*. Southeast Asia Program Publications, an imprint of Cornell University Press.

Breslawski, J. (2021). Armed groups and public health emergencies: A cross-country look at armed groups' responses to Covid-19. *Journal of Global Security Studies, 7*(1), ogab017. https://doi.org/10.1093/jogss/ogab017.

Broberg, M. (2020). Interpreting the UNFCCC's provisions on "mitigation" and "adaptation" in light of the Paris Agreement's provision on "loss and damage." *Climate Policy, 20*(5), 527–533. https://doi.org/10.1080/14693062.2020.1745744.

Brondizio, E. S., O'Brien, K., Bai, X., et al. (2016). Re-conceptualizing the anthropocene: A call for collaboration. *Global Environmental Change, 39*, 318–327. https://doi.org/10.1016/j.gloenvcha.2016.02.006.

Buhaug, H., Nordkvelle, J., Bernauer, T., et al. (2014). One effect to rule them all? A comment on climate and conflict. *Climatic Change, 127*(3–4), 391–397. https://doi.org/10.1007/s10584-014-1266-1.

Buhaug, H., & Von Uexkull, N. (2021). Vicious circles: Violence, vulnerability, and climate change. *Annual Review of Environment and Resources, 46*(1), 545–568. https://doi.org/10.1146/annurev-environ-012220-014708.

Callahan, M. P. (2007). *Political authority in Burma's ethnic minority states: Devolution, occupation and coexistence* (Vol. 31). Institute of Southeast Asian Studies.

Caris, C. C., & Reynolds, S. (2014). *ISIS governance in Syria* (Middle East Security Report 22). Institute for the Study of War. www.understandingwar.org/sites/default/files/ISIS_Governance.pdf.

Carnegie, A., Howe, K., Lichtenheld, A., & Mukhopadhyay, D. (2022). The effects of foreign aid on rebel governance: Evidence from a large-scale US aid program in Syria. *Economics & Politics, 34*(1), 41–66. https://doi.org/10.1111/ecpo.12178.

Cavanagh, C., & Benjaminsen, T. A. (2014). Virtual nature, violent accumulation: The "spectacular failure" of carbon offsetting at a Ugandan National Park. *Geoforum, 56*, 55–65. https://doi.org/10.1016/j.geoforum.2014.06.013.

Cemgil, C., & Hoffmann, C. (2016). The "Rojava revolution" in Syrian Kurdistan: A model of development for the Middle East? *IDS Bulletin, 47* (3), 53–76. https://doi.org/10.19088/1968-2016.144.

Chambers, J., & Kyed, H. M. (2024). *Bridging the gap in climate change financing to violent conflict affected areas* (DIIS Policy Brief). Danish Institute for International Studies. www.diis.dk/en/research/bridging-the-gap-in-climate-change-financing-to-violent-conflict-affected-areas.

Cheetah Conservation Fund. (2023, February 20). Cheetah conservation fund and Somaliland ministry of environment and climate change hold community based natural resource management workshop. *Cheetah Conservation Fund.*

Coggins, B. (2014). *Power politics and state formation in the twentieth century: The dynamics of recognition.* Cambridge University Press.

Collins, J. (2023, October 16). Armed groups as forest protectors in Colombia: A risky dynamic. *Sierra.* www.sierraclub.org/sierra/armed-groups-forest-amazon-protectors-colombia-risky-dynamic.

Conflict and Environment Observatory. (2021). *The environmental dimensions of the 2020 Nagorno-Karabakh conflict* (Publications, Law and Policy). https://ceobs.org/investigating-the-environmental-dimensions-of-the-nagorno-karabakh-conflict/.

Conrad, J. M., Greene, K. T., Walsh, J. I., & Whitaker, B. E. (2019). Rebel natural resource exploitation and conflict duration. *Journal of Conflict Resolution, 63*(3), 591–616. https://doi.org/10.1177/0022002718755853.

COP28. (2023). *COP28 declaration of climate, relief, recovery and peace.* https://aidmi.org/cop28-uae-climate-relief-recovery-and-peace-declaration/?gad_source=1&gclid=Cj0KCQiAv628BhC2ARIsAIJIiK9bVY9g_Ru2kfgI-gonezoz46oZmEsoOPryNSaxzpmaX1xHxLj-Gv0aAhGOEALw_wcB.

COP29. (2024, November 14). *COP29 Presidency launches Baku call on climate action for peace, relief, and recovery.* https://cop29.az/en/media-hub/news/cop29-presidency-launches-baku-call-on-climate-action-for-peace-relief-and-recovery.

Cullenward, D., Badgley, G., & Chay, F. (2023). Carbon offsets are incompatible with the Paris Agreement. *One Earth, 6*(9), 1085–1088. https://doi.org/10.1016/j.oneear.2023.08.014.

Cunningham, K. G., Huang, R., & Sawyer, K. M. (2021). Voting for militants: Rebel elections in civil war. *Journal of Conflict Resolution, 65*(1), 81–107. https://doi.org/10.1177/0022002720937750.

Cunningham, K. G., & Loyle, C. E. (2021). Introduction to the special feature on dynamic processes of rebel governance. *Journal of Conflict Resolution, 65* (1), 3–14. https://doi.org/10.1177/0022002720935153.

Cunningham, K. G., & Sawyer, K. (2019). Conflict negotiations and rebel leader selection. *Journal of Peace Research, 56*(5), 619–634. https://doi .org/10.1177/0022343319829689.

Daoudy, M. (2020). Water weaponization in the Syrian conflict: Strategies of domination and cooperation. *International Affairs, 96*(5), 1347–1366. https:// doi.org/10.1093/ia/iiaa131.

Daoudy, M. (2023). Climate change and regional instability in the Middle East. *Center for Preventive Action, Council on Foreign Relations* (pp. 1–30). Discussion Paper Series on Managing Global Disorder No. 14. www.cfr .org/report/climate-change-and-regional-instability-middle-east.

Darwish, M. (2023, November 10). The Olive Branch. *Atmos.* https://atmos .earth/overview-palestine-olive-trees-symbolism-ceasefire/.

Davies, S., Engström, G., Pettersson, T., & Öberg, M. (2024). Organized violence 1989–2023, and the prevalence of organized crime groups. *Journal of Peace Research, 61*(4), 673–693. https://doi.org/10.1177/ 00223433241262912.

Davies, S., Pettersson, T., & Öberg, M. (2023). Organized violence 1989–2022, and the return of conflict between states. *Journal of Peace Research, 60*(4), 691–708. https://doi.org/10.1177/00223433231185169.

De La Bourdonnaye, T. (2020). Greener insurgencies? Engaging non-State armed groups for the protection of the natural environment during non-international armed conflicts. *International Review of the Red Cross, 102*(914), 579–605. https://doi.org/10.1017/S1816383121000527.

Democracy Now. (2021, September 17). Climate colonialism: Why was occupied Western Sahara excluded from COP26 U.N. Summit in Scotland? *Democracy Now.* www.democracynow.org/2021/11/17/exclusion_ of_western_sahara_cop26_scotland.

Department of Environment and Natural Resources. (2003). *The Malitubog-Maridagao watershed rehabilitation subproject completion report* [Subproject Completion Report]. Republic of the Philippines.

Driscoll, J. (2015). *Warlords and coalition politics in post-Soviet states.* Cambridge University Press.

Druckman, J. N., & McGrath, M. C. (2019). The evidence for motivated reasoning in climate change preference formation. *Nature Climate Change, 9*(2), 111–119. https://doi.org/10.1038/s41558-018-0360-1.

Dubash, N. K. (2021). Varieties of climate governance: The emergence and functioning of climate institutions. *Environmental Politics, 30*(sup1), 1–25. https://doi.org/10.1080/09644016.2021.1979775.

Dura, T., Garner, A. J., Weiss, R., et al. (2021). Changing impacts of Alaska-Aleutian subduction zone tsunamis in California under future sea-level rise. *Nature Communications, 12*(1), 7119. https://doi.org/10.1038/s41467-021-27445-8.

Duyvesteyn, I. (2017). Rebels & legitimacy; An introduction. *Small Wars & Insurgencies, 28*(4–5), 669–685. https://doi.org/10.1080/09592318.2017.1322337.

Dzebo, A., & Stripple, J. (2015). Transnational adaptation governance: An emerging fourth era of adaptation. *Global Environmental Change, 35,* 423–435. https://doi.org/10.1016/j.gloenvcha.2015.10.006.

Ecosystem for Peace. (2024). *Common principles for effective climate finance and action for relief, recovery, and peace.* www.ecosystemforpeace.org/principles.

Eklund, L., Abdi, A., & Islar, M. (2017). From producers to consumers: The challenges and opportunities of agricultural development in Iraqi Kurdistan. *Land, 6*(2), 44. https://doi.org/10.3390/land6020044.

Ewing, J. J. (2009). *Converging peril: Climate change and conflict in the Southern Philippines.* www.jstor.org/stable/pdf/resrep17185.pdf.

Faye, W. (2006). *The Casamance separatism: From independence claim to resource logic* [M.A]. Naval Postgraduate School.

Fazal, T. M. (2011). *State death: The politics and geography of conquest, occupation, and annexation.* Princeton University Press. https://doi.org/10.1515/9781400841448.

Fergusson, L., Romero, D., & Vargas, J. F. (2014). The environmental impact of civil conflict: The deforestation effect of paramilitary expansion in Colombia. *SSRN Electronic Journal,* 1–41. https://doi.org/10.2139/ssrn.2516512.

Feuer, A. (2023). Environmental warfare tactics in irregular conflicts. *Perspectives on Politics, 21*(2), 533–549.

Fishbein, E., Brang, H. J., Awng, Z. M., & Hkawng, J. T. (2023, May 2). How the Kachin public overturned a rare earth mining project in KIO territory. *Frontier Myanmar.* www.frontiermyanmar.net/en/how-the-kachin-public-overturned-a-rare-earth-mining-project-in-kio-territory/.

Flanigan, S. T. (2008). Nonprofit service provision by insurgent organizations: The cases of Hizballah and the Tamil Tigers. *Studies in Conflict & Terrorism, 31*(6), 499–519. https://doi.org/10.1080/10576100802065103.

Florea, A. (2020). Rebel governance in de facto states. *European Journal of International Relations, 26*(4), 1004–1031. https://doi.org/10.1177/1354066120919481.

Florea, A., & Malejacq, R. (2023). The supply and demand of rebel governance. *International Studies Review, 26*(1), viae004. https://doi.org/10.1093/isr/viae004.

Furlan, M. (2020). Understanding governance by insurgent non-state actors: A multi-dimensional typology. *Civil Wars*, *22*(4), 478–511. https://doi.org/10.1080/13698249.2020.1785725.

Fuso Nerini, F., Sovacool, B., Hughes, N., et al. (2019). Connecting climate action with other sustainable development goals. *Nature Sustainability*, *2*(8), 674–680. https://doi.org/10.1038/s41893-019-0334-y.

Gharibian, T. (2021). *Advancing Regional Resilience in the Wake of War: A Proposed Resilience Framework for the Republic of Artsakh*. Georgetown University. https://repository.digital.georgetown.edu/handle/10822/1062769.

GebreMichael, T., Hundessa, T., & Hillma, J. (1992). *The effects of war on world heritage sites and protected areas in Ethiopia* (pp. 143–150).

Gerges, F. A. (2016). *ISIS: A history*. Princeton University Press.

Gilmore, E. A., & Buhaug, H. (2021). Climate mitigation policies and the potential pathways to conflict: Outlining a research agenda. *WIREs Climate Change*, *12*(5), e722. https://doi.org/10.1002/wcc.722.

Gleditsch, K. S., & Ward, M. D. (1999). A revised list of independent states since the congress of Vienna. *International Interactions*, *25*(4), 393–413. https://doi.org/10.1080/03050629908434958.

Gleditsch, N. P. (2012). Whither the weather? Climate change and conflict. *Journal of Peace Research*, *49*(1), 3–9. https://doi.org/10.1177/0022343311431288.

Gleditsch, N. P., Wallensteen, P., Eriksson, M., Sollenberg, M., & Strand, H. (2002). Armed conflict 1946–2001: A new dataset. *Journal of Peace Research*, *39*(5), 615–637. https://doi.org/10.1177/0022343302039005007.

Gupta, J. (2016). Climate change governance: History, future, and triple-loop learning? *WIREs Climate Change*, *7*(2), 192–210. https://doi.org/10.1002/wcc.388.

Hadden, J., & Bush, S. S. (2021). What's different about the environment? Environmental INGOs in comparative perspective. *Environmental Politics*, *30*(1–2), 202–223. https://doi.org/10.1080/09644016.2020.1799643.

Hale, T. (2018). The role of sub-state and non-state actors in international climate processes. *Londres*: *Chatham House*. www.chathamhouse.org/sites/default/files/publications/research/2018-11-28-non-state-sctors-climate-synthesis-hale-final.pdf.

Hatahet, S. (2019). *The political economy of the autonomous administration of north and east Syria* [Technical Report]. European University Institute. https://cadmus.eui.eu/handle/1814/65364.

Heger, L. L., & Jung, D. F. (2017). Negotiating with rebels: The effect of rebel service provision on conflict negotiations. *Journal of Conflict Resolution*, *61*(6), 1203–1229. https://doi.org/10.1177/0022002715603451.

Hendrix, C. S., & Glaser, S. M. (2011). Civil conflict and world fisheries, 1952–2004. *Journal of Peace Research, 48*(4), 481–495. https://doi.org/10.1177/0022343311399129

Hendrix, C. S., & Salehyan, I. (2012). Climate change, rainfall, and social conflict in Africa. *Journal of Peace Research, 49*(1), 35–50. https://doi.org/10.1177/0022343311426165.

Hofmann, C. (2012). *Reasoning with rebels: International NGOs' approaches to engaging armed groups* (No. RP 11/2012; SWP Research Paper). Stiftung Wissenschaft und Politik (SWP).

Hong, E. (2017). Scaling struggles over land and law: Autonomy, investment, and interlegality in Myanmar's borderlands. *Geoforum, 82*, 225–236. https://doi.org/10.1016/j.geoforum.2017.01.017.

Hossain, M. F., Huq, S., & Khan, M. R. (2021). The intractability of loss and damage issues in climate negotiations. *Soundings, 78*(78), 38–49. https://doi.org/10.3898/SOUN.78.02.2021.

Hsiang, S. M., Burke, M., & Miguel, E. (2013). Quantifying the influence of climate on human conflict. *Science, 341*(6151). https://doi.org/.10.1126/science.1235367.

Huang, R. (2016a). Rebel diplomacy in civil war. *International Security, 40*(4), 89–126. https://doi.org/10.1162/ISEC_a_00237.

Huang, R. (2016b). *The wartime origins of democratization: Civil war, rebel governance, and political regimes*. Cambridge University Press.

Huang, R., & Sullivan, P. L. (2021). Arms for education? External support and rebel social services. *Journal of Peace Research, 58*(4), 794–808.

Huddleston, R. J., & Hall, C. (2023). Secession and diplomacy: Playing the state, proving the nation. In R. D. Griffiths, A. Pavković, & P. Radan (eds.), *The Routledge handbook of self-determination and secession* (pp. 372–385). Routledge. www.taylorfrancis.com/chapters/edit/10.4324/9781003036593-30/secession-diplomacy-joseph-huddleston-caroline-hall.

Ide, T. (2023). *Catastrophes, confrontations, and constraints: How disasters shape the dynamics of armed conflicts*. MIT Press.

Ingram, H. J., Whiteside, C., & Winter, C. (2020). *The ISIS reader: Milestone texts of the Islamic State movement*. Oxford University Press.

IOM. (2014). *IOM contributions to progressively resolve displacement situations Compendium of activities and good practice*. International Organization for Migration. www.iom.int/sites/g/files/tmzbdl486/files/documents/IOM-Contributions-to-Progressively-Resolve-Displacement-Situations.pdf.

IPCC. (2022). Summary for Policymakers. In H.-O. Pörtner, D. C. Roberts, M. Tignor, et al. (eds.), *Climate change 2022: Impacts, adaptation, and vulnerability* (pp. 3–33). Contribution of Working Group II to the Sixth Assessment Report of the Intergovernmental Panel on Climate Change. Cambridge University Press. https://doi.org/10.1017/9781009325844.001.

IPCC. (2023a). *Climate change 2022 – Impacts, adaptation and vulnerability: Working group II contribution to the sixth assessment report of the intergovernmental panel on climate change* (1st ed.). Cambridge University Press. https://doi.org/10.1017/9781009325844.

IPCC. (2023b). Summary for policymakers. In Core Writing Team, H. Lee, & J. Romero (eds.), *Climate change 2023: Synthesis report* (pp. 1–34). Contribution of Working Groups I, II and III to the Sixth Assessment Report of the Intergovernmental Panel on Climate Change. IPCC. https://doi.org/10.59327/IPCC/AR6-9789291691647.001.

Jackson, A. (2018). *Life under the Taliban shadow government.* Overseas Development Institute.

Jackson, A., Weigand, F., Mayhew, L., & Bongard, P. (2023). *Climate adapatation in no-mans' land.* Research Report. ODI Global.

Jo, H. (2015). *Compliant rebels.* Cambridge University Press.

Jongerden, J. (2010). Dams and politics in Turkey: Utilizing water, developing conflict. *Middle East Policy, 17*(1), 137–143. https://doi.org/10.1111/j.1475-4967.2010.00432.x.

Jordan, A. J., Huitema, D., Hildén, M., et al. (2015). Emergence of polycentric climate governance and its future prospects. *Nature Climate Change, 5*(11), 977–982. https://doi.org/10.1038/nclimate2725.

Karagiannis, E. (2015). When the green gets greener: Political Islam's newly-found environmentalism. *Small Wars & Insurgencies, 26*(1), 181–201. https://doi.org/10.1080/09592318.2014.959768.

Kasfir, N., Frerks, G., & Terpstra, N. (2017). Introduction: Armed groups and multi-layered governance. *Civil Wars, 19*(3), 257–278. https://doi.org/10.1080/13698249.2017.1419611.

Keohane, R. O., & Victor, D. G. (2011). The regime complex for climate change. *Perspectives on Politics, 9*(1), 7–23. https://doi.org/10.1017/S1537592710004068.

KESAN. (n.d.). *Karen environmental and social action network.* https://kesan.asia/about-kesan/.

Keskitalo, E. C. H., & Preston, B. L. (eds.). (2019). *Research handbook on climate change adaptation policy.* Edward Elgar.

Khalaf, R. (2015). Governance without government in Syria: Civil society and state building during conflict. *Syria Studies, 7*(3), 37–72.

Kingsbury, D. (2018). *Western Sahara: International law, justice and natural resources.* Routledge.

Kita, S. M. (2019). Barriers or enablers? Chiefs, elite capture, disasters, and resettlement in rural Malawi. *Disasters, 43*(1), 135–156. https://doi.org/10.1111/disa.12295.

Kivimaa, P., Hildén, M., Huitema, D., Jordan, A., & Newig, J. (2017). Experiments in climate governance – A systematic review of research on energy and built environment transitions. *Journal of Cleaner Production, 169*, 17–29. https://doi.org/10.1016/j.jclepro.2017.01.027.

Knapp, M., & Jongerden, J. (2016). Communal democracy: The social contract and confederalism in Rojava. *Comparative Islamic Studies, 10*(1), 87–109. https://doi.org/10.1558/cis.29642.

Knutson, T. R., Chung, M. V., Vecchi, G., et al. (2021, March). *Climate change is probably increasing the intensity of tropical cyclones.* Science Brief Review. https://tyndall.ac.uk/wp-content/uploads/2023/10/ScienceBrief_Review_CYCLONES_Mar2021.pdf.

Koubi, V. (2019). Climate change and conflict. *Annual Review of Political Science, 22*(1), 343–360. https://doi.org/10.1146/annurev-polisci-050317-070830.

Koubi, V., Bernauer, T., Kalbhenn, A., & Spilker, G. (2012). Climate variability, economic growth, and civil conflict. *Journal of Peace Research, 49*(1), 113–127. https://doi.org/10.1177/0022343311427173.

Kuyper, J. W., Linnér, B., & Schroeder, H. (2018). Non-state actors in hybrid global climate governance: Justice, legitimacy, and effectiveness in a post-Paris era. *Wiley Interdisciplinary Reviews: Climate Change, 9*(1), e497.

Läderach, P., Ramirez-Villegas, J., Caroli, G., Sadoff, C., & Pacillo, G. (2021). Climate finance and peace – Tackling the climate and humanitarian crisis. *The Lancet Planetary Health, 5*(12), e856–e858. https://doi.org/10.1016/S2542-5196(21)00295-3.

Ledwidge, F. (2017). *Rebel law: Insurgents, courts and justice in modern conflict.* Hurst.

Leylekian, L. (2016). The Sarsang Reservoir in Upper Karabakh: Politicization of an environmental challenge in the framework of a territorial dispute. *Water Policy, 18*(2), 445–462. doi:10.2166/wp.2015.083.

Linke, A. M., & Ruether, B. (2021). Weather, wheat, and war: Security implications of climate variability for conflict in Syria. *Journal of Peace Research, 58*(1), 114–131. https://doi.org/10.1177/0022343320973070.

Linke, A. M., Witmer, F. D. W., O'Loughlin, J., McCabe, J. T., & Tir, J. (2018). Drought, local institutional contexts, and support for violence in Kenya. *Journal of Conflict Resolution, 62*(7), 1544–1578. https://doi.org/10.1177/0022002717698018.

Loyle, C. E. (2021). Rebel justice during armed conflict. *Journal of Conflict Resolution*, *65*(1), 108–134. https://doi.org/10.1177/0022002720939299.

Loyle, C. E., Braithwaite, J. M., Cunningham, K. G., et al. (2022). Revolt and rule: Learning about governance from rebel groups. *International Interactions*, *24*(4), viac043.

Loyle, C. E., Cunningham, K. G., Huang, R., & Jung, D. F. (2023). New directions in rebel governance research. *Perspectives on Politics*, *21*(1), 264–276. https://doi.org/10.1017/S1537592721001985.

Lujala, P. (2010). The spoils of nature: Armed civil conflict and rebel access to natural resources. *Journal of Peace Research*, *47*(1), 15–28. https://doi.org/10.1177/0022343309350015.

Malejacq, R. (2019). *Warlord survival: The delusion of state building in Afghanistan*. Cornell University Press.

Mampilly, Z. C. (2011). *Rebel rulers: Insurgent governance and civilian life during war*. Cornell University Press.

Mampilly, Z. C. (2015). Performing the nation-state: Rebel governance and symbolic processes. In A. Arjona, N. Kasfir, & Z. C. Mampilly (eds.), *Rebel Governance in civil war* (pp. 74–97). Cambridge University Press.

Marks, Z. (2019). Rebel resource strategies in civil war: Revisiting diamonds in Sierra Leone. *Political Geography*, *75*, 102059. https://doi.org/10.1016/j.polgeo.2019.102059.

Martínez, J. C., & Eng, B. (2018). Stifling stateness: The Assad regime's campaign against rebel governance. *Security Dialogue*, *49*(4), 235–253. https://doi.org/10.1177/0967010618768622.

Matfess, H. (2022). The need for a cooperation framework between non-state armed groups and humanitarians. Center for Strategic and International Studies. www.csis.org/analysis/need-cooperation-framework-between-non-state-armed-groups-and-humanitarians.

Mbade, S. A. (2018). Effects of armed conflict on Lower Casamance National Park environment in the southwestern Senegal. *International Journal of Development Research*, *8*(6), 21285–21290.

McGregor, A. (2010). Green and Redd? Towards a political ecology of deforestation in Aceh, Indonesia. *Human Geography*, *3*(2), 21–34. https://doi.org/10.1177/194277861000300202.

McWeeney, M., Cunningham, K. G., & Bauer, L. (2023). Rebel actors and legitimacy building. *International Politics*. https://doi.org/10.1057/s41311-023-00493-1.

Medina, L., Krendelsberger, A., Renkamp, T., et al. (2023). *Community voices on climate, peace and security: Senegal* (CGIAR FOCUS Climate Security). https://hdl.handle.net/10568/137319.

Medina, L., Pacillo, G., Läderach, P., Sieber, S., & Bonatti, M. (2025). Adapting to climate change under threats of violence: A comparative institutional analysis of incentives for conflict and collaboration. *Current Research in Environmental Sustainability*, *9*, 100276. https://doi.org/10.1016/j.crsust.2024.100276.

Mena, R., & Hilhorst, D. (2021). The (im)possibilities of disaster risk reduction in the context of high-intensity conflict: The case of Afghanistan. *Environmental Hazards*, *20*(2), 188–208. https://doi.org/10.1080/17477891.2020.1771250.

Middle East | Action against Hunger – Global Hunger Relief. (n.d.). Action against Hunger. Retrieved October 10, 2023, from www.actionagainsthunger.org/location/middle-east/.

Minttu, R. (2023, July 17). Water, power, and life at risk: Tishrin dam. *Medium.* https://ronkaminttu.medium.com/water-power-and-life-at-risk-tishrin-dam-c296cd369574.

Muller, M. (2018). Cape Town's drought: Don't blame climate change. *Nature*, *559*(7713), 174–176. https://doi.org/10.1038/d41586-018-05649-1.

Munive, J., & Stepputat, F. (2023). *How non-state armed groups engage in environmental protection* (DIIS Policy Brief). www.diis.dk/en/research/how-non-state-armed-groups-engage-in-environmental-protection.

Musah-Surugu, I. J., Ahenkan, A., & Bawole, J. N. (2019). Too weak to lead: Motivation, agenda setting and constraints of local government to implement decentralized climate change adaptation policy in Ghana. *Environment, Development and Sustainability*, *21*(2), 587–607. https://doi.org/10.1007/s10668-017-0049-z.

Nalau, J., Preston, B. L., & Maloney, M. C. (2015). Is adaptation a local responsibility? *Environmental Science & Policy*, *48*, 89–98. https://doi.org/10.1016/j.envsci.2014.12.011.

Nalbo, D. (2019). *The "local" and peacebuilding: A study on community forests in Myanmar's Kachin State.* George Mason University.

Okereke, C., Bulkeley, H., & Schroeder, H. (2009). Conceptualizing climate governance beyond the international regime. *Global Environmental Politics*, *9*(1), 58–78. https://doi.org/10.1162/glep.2009.9.1.58.

Okumu, W. (2003). Humanitarian international NGOs and African conflicts. *International Peacekeeping*, *10*(1), 120–137. https://doi.org/10.1080/714002390.

Omer, M. A. (2024). Climate variability and livelihood in Somaliland: A review of the impacts, gaps, and ways forward. *Cogent Social Sciences*, *10*(1), 2299108. https://doi.org/10.1080/23311886.2023.2299108.

ONLF. (n.d.). *Political programme of the Ogaden National Liberation Front (ONLF)*. Geneva Call. http://theirwords.org/media/transfer/doc/et_onlf_02-3c7a7281a188e37a9c88003e82188845.pdf.

Oo, Z., & Min, W. (2007). *Assessing Burma's ceasefire accords* (Vol. 39). Institute of Southeast Asian Studies.

Otto, F. E. L. (2017). Attribution of weather and climate events. *Annual Review of Environment and Resources, 42*(1), 627–646. https://doi.org/10.1146/annurev-environ-102016-060847.

Owen, G. (2020). What makes climate change adaptation effective? A systematic review of the literature. *Global Environmental Change, 62,* 102071. https://doi.org/10.1016/j.gloenvcha.2020.102071.

Özçelik, B. (2020). Explaining the Kurdish Democratic Union Party's self-governance practices in Northern Syria, 2012–18. *Government and Opposition, 55*(4), 690–710. https://doi.org/10.1017/gov.2019.1.

Parkinson, S. E. (2013). Organizing rebellion: Rethinking high-risk mobilization and social networks in war. *American Political Science Review, 107*(3), 418–432. https://doi.org/10.1017/S0003055413000208.

Paul, A., Roth, R., & Sein Twa, S. P. (2023). Conservation for self-determination: Salween peace park as an Indigenous Karen conservation initiative. *AlterNative: An International Journal of Indigenous Peoples, 19* (2), 271–282. https://doi.org/10.1177/11771801231169044.

Pearce, F. (2020). *Amid tensions in Myanmar, an indigenous park of peace is born.* https://e360.yale.edu/features/amid-tensions-in-myanmar-an-indigenous-park-of-peace-is-born.

Peluso, N. L., & Vandergeest, P. (2011). Political ecologies of war and forests: Counterinsurgencies and the making of national natures. *Annals of the Association of American Geographers, 101*(3), 587–608. https://doi.org/10.1080/00045608.2011.560064.

Persson, Å. (2019). Global adaptation governance: An emerging but contested domain. *WIREs Climate Change, 10*(6), 1–18. https://doi.org/10.1002/wcc.618.

Philipps, D., Khan, A., & Schmitt, E. (2022, January 20). A dam in Syria was on a "no-strike" list. The U.S. bombed it anyway. *The New York Times.* www.nytimes.com/2022/01/20/us/airstrike-us-isis-dam.html.

Phillips, S. G. (2020). *When there was no aid: War and peace in Somaliland.* Cornell University Press. https://doi.org/10.1515/9781501747168.

Podder, S. (2017). Understanding the legitimacy of armed groups: A relational perspective. *Small Wars & Insurgencies, 28*(4–5), 686–708. https://doi.org/10.1080/09592318.2017.1322333.

Porges, M. (2020). Environmental challenges and local strategies in Western Sahara. *Forced Migration Review, 64*. www.fmreview.org/issue64/porges.

Pörtner, H.-O., Roberts, D. C., Poloczanska, E. S., et al. (eds.). (2023). IPCC, 2022: Summary for policymakers. In *Climate change 2022 – Impacts, adaptation and vulnerability: Working group II contribution to the sixth assessment report of the intergovernmental panel on climate change* (pp. 3–33). Cambridge University Press.

Raleigh, C., Linke, A., Barrett, S., & Kazemi, E. (2024). Climate finance and conflict: Adaptation amid instability. *The Lancet Planetary Health, 8*(1), e51–e60. https://doi.org/10.1016/S2542-5196(23)00256-5.

Reed, J. T., & Raschke, D. (2010). *The ETIM: China's Islamic militants and the global terrorist threat*. Praeger.

Republic of Somaliland. (2017). *Republic of Somaliland ministry of environment and rural development strategic plan (2017–2021)*. www.google.com/url?q=https://faolex.fao.org/docs/pdf/som209311.pdf&sa=D&source=docs&ust=1717518415189814&usg=AOvVaw2A5C5_fIXe-kbYNAhUgTQp.

Republic of Somaliland. (2018). *Somaliland environmental management law no. 79 of 2018*. www.fao.org/faolex/results/details/en/c/LEX-FAOC208680/.

Revkin, M. R. (2020). What explains taxation by resource-rich rebels? Evidence from the Islamic State in Syria. *The Journal of Politics, 82*(2), 757–764. https://doi.org/10.1086/706597.

Rubin, M. A. (2020). Rebel territorial control and civilian collective action in civil war: Evidence from the communist insurgency in the Philippines. *Journal of Conflict Resolution, 64*(2–3), 459–489. https://doi.org/10.1177/0022002719863844.

Sadan, M. (2015). Ongoing conflict in the Kachin state. *Southeast Asian Affairs, 2015*(1), 246–259.

SADR Office of the Prime Minister. (2021). *First indicative nationally determined contribution*. https://vest-sahara.s3.amazonaws.com/wsrw/feature-images/File/438/61892f66d5311_SADR-NDC_2021.pdf.

Salehyan, I., Gleditsch, K. S., & Cunningham, D. E. (2011). Explaining external support for insurgent groups. *International Organization, 65*(4), 709–744.

Sapiains, R., Ibarra, C., Jiménez, G., et al. (2021). Exploring the contours of climate governance: An interdisciplinary systematic literature review from a southern perspective. *Environmental Policy and Governance, 31*(1), 46–59. https://doi.org/10.1002/eet.1912.

Sary, G. (2016). *Kurdish self-governance in Syria: Survival and ambition* (Middle East and North Africa Programme). The Royal Institute of International Affairs. www.chathamhouse.org/sites/default/files/publications/research/2016-09-15-kurdish-self-governance-syria-sary-embargoed.pdf.

Scartozzi, C. M. (2024). Conflict sensitive climate finance: Lessons from the green climate fund. *Climate Policy, 24*(3), 297–313. https://doi.org/10.1080/14693062.2023.2212640.

Scartozzi, C. M., Savilli, A., Meddings, G., & Läderach, P. (2023). *Integrated climate security programming in climate finance: An analysis of multilateral climate funds* (CGIAR FOCUS Climate Security & CGIAR ClimBeR).

Schievels, J. J., & Colley, T. (2021). Explaining rebel-state collaboration in insurgency: Keep your friends close but your enemies closer. *Small Wars & Insurgencies, 32*(8), 1332–1361. https://doi.org/10.1080/09592318.2020.1827847.

Schipper, E. L. F., Dubash, N. K., & Mulugetta, Y. (2021). Climate change research and the search for solutions: Rethinking interdisciplinarity. *Climatic Change, 168*(3–4), 18. https://doi.org/10.1007/s10584-021-03237-3.

Schoon, E. W. (2017). Building legitimacy: Interactional dynamics and the popular evaluation of the Kurdistan Workers' Party (PKK) in Turkey. *Small Wars & Insurgencies, 28*(4–5), 734–754. https://doi.org/10.1080/09592318.2017.1323407.

Sen, A. K. (2023). *Kurdish official lists ISIS and climate change as top threats.* United States Institute of Peace.

Seymour, L. J. M. (2017). Legitimacy and the politics of recognition in Kosovo. *Small Wars & Insurgencies, 28*(4–5), 817–838. https://doi.org/10.1080/09592318.2017.1322335.

Shackleton, S., Ziervogel, G., Sallu, S., Gill, T., & Tschakert, P. (2015). Why is socially-just climate change adaptation in sub-Saharan Africa so challenging? A review of barriers identified from empirical cases. *WIREs Climate Change, 6*(3), 321–344. https://doi.org/10.1002/wcc.335.

Sharrow, Steven H. (2007). Natural resource management on the other side of the world: The Nagorno Karabakh Republic. *Rangelands 29*(1), 11–16.

Siddiqi, A. (2018). Disasters in conflict areas: Finding the politics. *Disasters, 42* (S2), S161–S172. https://doi.org/10.1111/disa.12302.

Sifuma, J. (2016). *Water infrastructure development for resilience in Somaliland program* [Environmental and Social Management Framework (ESMF) Summary]. African Development Bank Group. www.afdb.org/filead min/uploads/afdb/Documents/Environmental-and-Social-Assessments/Somaliland_-_Water_infrastructure_development_for_resilience_in_Somaliland_program_%E2%80%93_ESMF_Summary.pdf.

Simpson, N. P., Andrews, T. M., Krönke, M., et al. (2021). Climate change literacy in Africa. *Nature Climate Change, 11*(11), 937–944. https://doi.org/10.1038/s41558-021-01171-x.

Sinolinding, H. M., Porciuncula, F. L., & Corpuz, O. S. (2013). Conservation of Ligawasan Marsh in Mindanao, Philippines, through an indigenous knowledge system: Climate change mitigation and disaster risk management. In W. Leal Filho (ed.), *Climate change and disaster risk management* (pp. 615–626). Springer. https://doi.org/10.1007/978-3-642-31110-9_40.

Sitati, A., Joe, E., Pentz, B., et al. (2021). Climate change adaptation in conflict-affected countries: A systematic assessment of evidence. *Discover Sustainability, 2*(1), 42.

Smith, J. J. (2018). The taking of the Sahara: The role of natural resources in the continuing occupation of Western Sahara. In *Western Sahara: International Law, Justice and Natural Resources by Damien Kingsbury* (pp. 23–43).

Sottimano, A., & Samman, N. (2022, February 24). Syria has a water crisis: And it's not going away. *Atlantic Council.* www.atlanticcouncil.org/blogs/mena source/syria-has-a-water-crisis-and-its-not-going-away/.

Staniland, P. (2012). Organizing insurgency: Networks, resources, and rebellion in South Asia. *International Security, 37*(1), 142–177. https://doi.org/10.1162/ISEC_a_00091.

Stanton, J. A. (2020). Rebel groups, international humanitarian law, and civil war outcomes in the Post-Cold War era. *International Organization, 74*(3), 523–559. https://doi.org/10.1017/S0020818320000090.

Stewart, M. A. (2018). Civil war as state-making: Strategic governance in civil war. *International Organization, 72*(1), 205–226.

Stewart, M. A. (2021). *Governing for revolution: Social transformation in civil war.* Cambridge University Press.

Stewart, R. B., Oppenheimer, M., & Rudyk, B. (2013). A new strategy for global climate protection. *Climatic Change, 120*(1–2), 1–12. https://doi.org/10.1007/s10584-013-0790-8.

Tapscott, R. (2023). Vigilantes and the state: Understanding violence through a security assemblages approach. *Perspectives on Politics, 21*(1), 209–224. Cambridge Core. https://doi.org/10.1017/S1537592721001134.

Taya, S. L. (2007). The political strategies of the Moro Islamic Liberation Front for self-determination in the Philippines. *Intellectual Discourse, 15*(1). https://doi.org/10.31436/id.v15i1.61.

Taya, S. L. (2009). *Strategies and tactics of the Moro Islamic Liberation Front (MILF) in the Southern Philippines* (1st ed). Universiti Utara Malaysia Press.

Terpstra, N. (2020). Rebel governance, rebel legitimacy, and external intervention: Assessing three phases of Taliban rule in Afghanistan. *Small Wars & Insurgencies, 31*(6), 1143–1173. https://doi.org/10.1080/09592318.2020.1757916.

Terpstra, N., & Frerks, G. (2016). *Rebel governance: The cases of the LTTE and Hezbollah* [Policy Brief]. Gerda Henkel Stiftung. https://research-portal.uu.nl/ws/portalfiles/portal/24075571/Rebel_Governance_the_cases_of_the_LTTE_and_Hezbollah.pdf.

Terpstra, N., & Frerks, G. (2017). Rebel governance and legitimacy: Understanding the impact of rebel legitimation on civilian compliance with the LTTE rule. *Civil Wars*, *19*(3), 279–307. https://doi.org/10.1080/13698249.2017.1393265.

Tilly, C. (1985). War making and state making as organized crime. In P. B. Evans, D. Rueschemeyer, & T. Skocpol (eds.), *Bringing the State BackIn* (1st ed., pp. 169–191). Cambridge University Press. https://doi.org/10.1017/CBO9780511628283.008.

UDCP. (n.d.). *UDCP actor profile OPM*. https://ucdp.uu.se/actor/266.

ULMWP. (2021). *Green state vision – West Papua aims to be earth's first Green State*. https://greenstatevision.info/.

UNDP. (2016). *Quarterly progress report: Enhancing climate resilience of the vulnerable communities and ecosystems in Somalia (Atlas ID: 00084974)*. United Nations Development Program. www.undp.org/sites/g/files/zskgke326/files/migration/so/32ec21e0f917e76021d367c339c54c97baaead98c223b2405acd4e6cd671ed8b.pdf.

UNDP Somalia. (2016, January 6). Empowering pastoralist communities to manage climate change. *Medium*. https://undpsom.medium.com/empowering-pastoralist-communities-to-manage-climate-change-470ef1fb5f75.

UNHCR. (2012). *Global overview 2011: People internally displaced by conflict and violence – The Philippines* [Global Overview]. United Nations High Commissioner for Refugees. www.refworld.org/reference/annualreport/idmc/2012/en/85805.

UNHCR. (2024, August 29). *Syria situation*. UNHCR Global Focus. https://reporting.unhcr.org/operational/situations/syria-situation#:~:text=At%20the%20end%20of%202023,and%20asylum%2Dseekers%20in%20Syria.

United Nations Development Programme. (2021). *Climate finance for sustaining peace: Making climate finance work for conflict-affected and fragile contexts*. www.undp.org/publications/climate-finance-sustaining-peace-making-climate-finance-work-conflict-affected-and-fragile-contexts.

United States Institute of Peace. (n.d.). *What is UNSCR 1325? An explanation of the landmark resolution on women, peace and security*. www.usip.org/gender_peacebuilding/about_UNSCR_1325#:~:text=UNSCR%201325%20affirms%20that%20peace,the%20forging%20of%20lasting%20peace.

Van Baalen, S. (2021). Local elites, civil resistance, and the responsiveness of rebel governance in Côte d'Ivoire. *Journal of Peace Research, 58*(5), 930–944. https://doi.org/10.1177/0022343320965675.

Van Der Heijden, J. (2019). Studying urban climate governance: Where to begin, what to look for, and how to make a meaningful contribution to scholarship and practice. *Earth System Governance, 1*, 100005. https://doi.org/10.1016/j.esg.2019.100005.

Van Etten, J., Jongerden, J., De Vos, H. J., Klaasse, A., & Van Hoeve, E. C. E. (2008). Environmental destruction as a counterinsurgency strategy in the Kurdistan region of Turkey. *Geoforum, 39*(5), 1786–1797. https://doi.org/10.1016/j.geoforum.2008.05.001.

Venner, K., García-Lamarca, M., & Olazabal, M. (2024). The multi-scalar inequities of climate adaptation finance: A critical review. *Current Climate Change Reports.* https://doi.org/10.1007/s40641-024-00195-7.

VOA News. (2018, July 3). Eco-Jihadism: Somali terrorist group bans plastic bags. *VOA News.* www.voanews.com/a/somalia-terrorist-group-al-shabab-bans-plastic-bags/4465508.html.

von Lossow, T. (2016). The rebirth of water as a weapon: IS in Syria and Iraq. *The International Spectator, 51*(3), 82–99. https://doi.org/10.1080/03932729.2016.1213063.

Von Uexkull, N., & Buhaug, H. (2021). Security implications of climate change: A decade of scientific progress. *Journal of Peace Research, 58*(1), 3–17. https://doi.org/10.1177/0022343320984210.

Waisová, Š. (2017). *Environmental cooperation as a tool for conflict transformation and resolution.* Lexington Books.

Walch, C. (2014). Collaboration or obstruction? Rebel group behavior during natural disaster relief in the Philippines. *Political Geography, 43*, 40–50. https://doi.org/10.1016/j.polgeo.2014.09.007.

Walch, C. (2018). Disaster risk reduction amidst armed conflict: Informal institutions, rebel groups, and wartime political orders. *Disasters, 42*(S2), S239–S264. https://doi.org/10.1111/disa.12309.

Walsh, J. I., Conrad, J. M., Whitaker, B. E., & Hudak, K. M. (2018). Funding rebellion: The rebel contraband dataset. *Journal of Peace Research, 55*(5), 699–707.

Weinstein, J. M. (2006). *Inside rebellion: The politics of insurgent violence.* Cambridge University Press.

West, T. A. P., Börner, J., Sills, E. O., & Kontoleon, A. (2020). Overstated carbon emission reductions from voluntary REDD+ projects in the Brazilian Amazon. *Proceedings of the National Academy of Sciences, 117*(39), 24188–24194. https://doi.org/10.1073/pnas.2004334117.

Western Sahara Resource Watch. (2021, October 6). Report: Morocco uses green energy to embellish its occupation. *Western Sahara Resource Watch.* https://wsrw.org/en/news/report-morocco-uses-green-energy-to-embellish-its-occupation.

Williams, C. G. (2024). A forest peacebuilding mechanism for the world's most fragile states. *Environmental Science & Policy, 157,* 103749. https://doi.org/10.1016/j.envsci.2024.103749.

Williams, E. (2020). Attributing blame? – Climate accountability and the uneven landscape of impacts, emissions, and finances. *Climatic Change, 161*(2), 273–290. https://doi.org/10.1007/s10584-019-02620-5.

Wong, C. (2022). Climate finance and the peace dividend, articulating the co-benefits argument. In C. Cash, & L. A. Swatuk (eds.), *The political economy of climate finance: Lessons from international development* (pp. 205–231). Springer. https://doi.org/10.1007/978-3-031-12619-2_9.

Wood, R. M., & Sullivan, C. (2015). Doing harm by doing good? The negative externalities of humanitarian aid provision during civil conflict. *The Journal of Politics, 77*(3), 736–748. https://doi.org/10.1086/681239.

Woods, K. (2011). The commercalisation of counterinsurgency: Battlefield enemies, business bedfellows in Kachin state. In Mandy Sagan (ed.), *War and Peace in the Borderlands of Myanmar: The Kachin ceasefire, 1994–2011* (pp. 114–145). NIAS Press.

Woods, K. (2019). *Natural resource governance reform and the peace process in Myanmar* (Forest Trends: Forest Policy Trade and Finance Initiative Report). www.researchgate.net/publication/339166058_Natural_Resource_Governance_Reform_and_Myanmar%27s_Peace_Process.

World Bank. (2007). *Mindanao trust fund reconstruction and development program, annual report 2006* [Annual Report]. The International Bank for Reconstruction and Development/World Bank. https://documents.worldbank.org/pt/publication/documents-reports/documentdetail/416441468333586711/mindanao-trust-fund-reconstruction-and-development-program-annual-report-2006.

Worrall, J. (2018). (Re-) emergent orders: Understanding the negotiation (s) of rebel governance. In I. Duyvesteyn (ed.), *Rebels and legitimacy: Processes and practices* (pp. 41–65). Routledge.

Wrathall, D. J., Devine, J., Aguilar-González, B., et al. (2020). The impacts of cocaine-trafficking on conservation governance in Central America. *Global Environmental Change, 63,* 102098. https://doi.org/10.1016/j.gloenvcha.2020.102098.

Yadin, S. (2023). *Fighting climate change through shaming.* Cambridge University Press.

Zanatta, L., & Alvi M. (2024). Assessing water (ir)rationality in Nagorno-Karabakh. In Anja Mihr & Chiara Pierobon (eds.), *Polarization, shifting borders and liquid governance: Studies on transformation and development in the OSCE region* (pp. 79–97). Springer Nature Switzerland, doi:10.1007/978-3-031-44584-2_5.

Acknowledgment

The authors thank and acknowledge the support of our research team: Pedro Abelin, Leo Bauer, Liv Bauer, Madie Fleishman, Harriet Goers, Juan David Gelvez, Will Herens, Nat Hill, Sloan Lansdale, Badradin Mohammed, Greg O'Connell, Brian Wendelgass, Taylor Vincent, and Fatima Younis, our editors Aseem Prakash and Jennifer Hadden, and Kelly Morrison and Jack Swab. This work was supported by the Minerva Initiative (*FA9550-22-1-0259*). Special thanks to David Cunningham for his support during the writing of this book.

Organizational Response to Climate Change

Aseem Prakash
University of Washington

Aseem Prakash is Professor of Political Science, the Walker Family Professor for the College of Arts and Sciences, and the Founding Director of the Center for Environmental Politics at University of Washington, Seattle. His recent awards include the American Political Science Association's 2020 Elinor Ostrom Career Achievement Award in recognition of "lifetime contribution to the study of science, technology, and environmental politics," the International Studies Association's 2019 Distinguished International Political Economy Scholar Award that recognizes "outstanding senior scholars whose influence and path-breaking intellectual work will continue to impact the field for years to come," and the European Consortium for Political Research Standing Group on Regulatory Governance's 2018 Regulatory Studies Development Award that recognizes a senior scholar who has made notable "contributions to the field of regulatory governance."

Jennifer Hadden
University of Maryland

Jennifer Hadden is Associate Professor in the Department of Government and Politics at the University of Maryland. She conducts research in international relations, environmental politics, network analysis, nonstate actors, and social movements. Her research has been published in various journals, including the *British Journal of Political Science, International Studies Quarterly, Global Environmental Politics, Environmental Politics,* and *Mobilization.* Dr. Hadden's award-winning book, *Networks in Contention: The Divisive Politics of Global Climate Change,* was published by Cambridge University Press in 2015. Her research has been supported by a Fulbright Fellowship, as well as grants from the National Science Foundation, the National Socio-Environmental Synthesis Center, and others. She held an International Affairs Fellowship from the Council on Foreign Relations for the 2015–16 academic year, supporting work on the Paris Climate Conference in the Office of the Special Envoy for Climate Change at the US Department of State.

David Konisky
Indiana University

David Konisky is Professor at the Paul H. O'Neill School of Public and Environmental Affairs, Indiana University, Bloomington. His research focuses on US environmental and energy policy, with particular emphasis on regulation, federalism and state politics, public opinion, and environmental justice. His research has been published in various journals, including *the American Journal of Political Science, Climatic Change,* the *Journal of Politics, Nature Energy,* and *Public Opinion Quarterly.* He has authored or edited six books on environmental politics and policy, including *Fifty Years at the U.S. Environmental Protection Agency: Progress, Retrenchment and Opportunities* (Rowman & Littlefield, 2020, with Jim Barnes and John D. Graham), *Failed Promises: Evaluating the Federal Government's Response to Environmental Justice* (MIT, 2015), and *Cheap and Clean: How Americans Think about Energy in the Age of Global Warming* (MIT, 2014, with Steve Ansolabehere). Konisky's research has been funded by the National Science Foundation, the Russell Sage Foundation, and the Alfred P. Sloan Foundation. Konisky is currently coeditor of *Environmental Politics.*

Matthew Potoski

UC Santa Barbara

Matthew Potoski is a Professor at UCSB's Bren School of Environmental Science and Management. He currently teaches courses on corporate environmental management, and his research focuses on management, voluntary environmental programs, and public policy. His research has appeared in business journals such as *Strategic Management Journal, Business Strategy and the Environment,* and the *Journal of Cleaner Production,* as well as public policy and management journals such as *Public Administration Review* and the *Journal of Policy Analysis and Management.* He coauthored *The Voluntary Environmentalists* (Cambridge, 2006) and *Complex Contracting* (Cambridge, 2014; the winner of the 2014 Best Book Award, American Society for Public Administration, Section on Public Administration Research) and was coeditor of *Voluntary Programs* (MIT, 2009). Professor Potoski is currently coeditor of the *Journal of Policy Analysis and Management* and the *International Public Management Journal.*

About the Series

How are governments, businesses, and nonprofits responding to the climate challenge in terms of what they do, how they function, and how they govern themselves? This series seeks to understand why and how they make these choices and with what consequence for the organization and the eco-system within which it functions.

For EU product safety concerns, contact us at Calle de José Abascal, 56–1°,
28003 Madrid, Spain or eugpsr@cambridge.org.

www.ingramcontent.com/pod-product-compliance
Ingram Content Group UK Ltd.
Pitfield, Milton Keynes, MK11 3LW, UK
UKHW021930220625

459989UK00014B/87